PUBLIC POLICY ISSUES IN RESOURCE MANAGEMENT

James A. Crutchfield and Robert H. Pealy, Editors

PUBLIC POLICY ISSUES IN RESOURCE MANAGEMENT

1. *The Fisheries: Problems in Resource Management*, edited by James A. Crutchfield
2. *Water Resources Management and Public Policy*, edited by Thomas H. Campbell and Robert O. Sylvester
3. *Weather Modification: Science and Public Policy*, edited by Robert G. Fleagle
4. *World Fisheries Policy: Multidisciplinary Views*, edited by Brian J. Rothschild
5. *Ocean Resources and Public Policy*, edited by T. Saunders English

Ocean Resources
and Public Policy

EDITED BY T. SAUNDERS ENGLISH

Seattle · UNIVERSITY OF WASHINGTON PRESS · London

Library of Congress Cataloging in Publication Data
Main entry under title:

Ocean resources and public policy.

(Public policy issues in resource management, v. 5)
Based on a series of interdisciplinary seminars held
at the University of Washington during the 1967–68
academic year.
Includes bibliographies.
1. Marine resources and state—Addresses, essays,
lectures. I. English, Thomas Saunders, 1928–   ed.
II. Series.
GC1015.032   333.9'164   77-103298
ISBN 0-295-95260-1

Preface

THIS volume on ocean resources and public policy contains eleven papers which evolved from a series of sixteen interdisciplinary seminars organized by Professor Richard H. Fleming. The consideration of ocean resources seemed timely and appropriate for inclusion in a series of seminars on natural resources and public policy organized under the Graduate School of Public Affairs of the University of Washington. The earlier publications in this series have dealt with fisheries management, water resources management, and weather modification.

The natural resources seminars are given by established experts in their individual disciplines. The audience is primarily graduate students and faculty, with invited representation from government and industry. The professional distinction of the speakers and the breadth of interest in ocean resources and public policy resulted in attendance from such diverse campus groups as the Graduate School of Public Affairs, the School of Law, the Departments of Atmospheric Sciences, Botany, Chemistry, Economics, Engineering, Fisheries, Forestry, Geography, Geology, Oceanography, Physics, and Zoology.

The seminar, held during the 1967/68 academic year, was planned so that the major resource which the ocean represents could be considered broadly by a gathering of natural and social scientists, while attempting to include the current thinking of industrialists, bureaucrats, and generalists (politicians). The issues of public policy that were raised included, among others, national security, economic growth, environmental quality, resource development, education, and human welfare. It became clear that new and challenging scientific questions gained further dimensions when considered within a socioeconomic framework. New perspectives emerged, for faculty as well as students, of the complex interactions within fields of study and between academic disciplines. The fallacies inherent in considering the sea

v

67169

apart from the land, or the air, became apparent. The breadth of alternative uses for the ocean suggested national and international aspects to any allocation of resources among claimants.

The papers in this volume were chosen for their continuing interest to the intelligent layman. The value of the collection lies in the contrast of viewpoints, which the passage of time will not obscure. The illumination of interfaces between disciplines is often explicit and always implicit. Five contributors cover the classic four fields within oceanography—biology, chemistry, geology, and physics. Fisheries and energy from oil are considered. Then the ocean is viewed by an economist, a lawyer, and a high government official.

The contributors are experts in their fields whose considerations of the ocean from different viewpoints leave some apparent and some real contradictions—which time and new knowledge may resolve.

T. Saunders English
Department of Oceanography,
University of Washington

Contents

Man and the Ocean   3
    RICHARD H. FLEMING

Coastlines and Continental Shelves:
    Geological History and Characteristics   11
    JOE S. CREAGER

Physical Oceanography of Estuaries   25
    MAURICE RATTRAY, JR.

Global Distribution of Organic Production
    in the Oceans   38
    KARL BANSE

The Marine Food Chain and Its Efficiency   49
    J. D. H. STRICKLAND

Food from the Sea and Public Policy   64
    WILBERT MCLEOD CHAPMAN

The Ocean Margins   76
    JOHN D. ISAACS

Energy from the Oceans   94
    RICHARD A. GEYER

Resources from the Sea   105
      JAMES A. CRUTCHFIELD

*An International Legal-Political Framework for Exploring
and Exploiting the Mineral Resources Underlying the
High Seas: The Recommendations of the Commission
on Marine Science, Engineering and Resources*   134
      CARL A. AUERBACH

New Machinery for Policy Planning in Marine Sciences   168
      EDWARD WENK, JR.

Index   179

OCEAN RESOURCES AND PUBLIC POLICY

Man and the Ocean

RICHARD H. FLEMING

It is inevitable that mankind will make greater use of the ocean. Depletion of nonrenewable terrestrial resources, in the face of rapidly rising demands for raw materials and energy, will force us to seek alternative sources of or equivalent replacements for these essential commodities. The world-wide demands for renewable resources such as food, timber, and fresh water are already in excess of the rates of supply, at least on regional or national bases. To support services and functions such as transportation and waste disposal, society has many needs, other than for supplies of various materials and for energy, that are becoming more and more pressing as a result of technological developments. The primary question is to what extent the ocean can replace or supplement these resources that in the past have been primarily provided by the land.

WHAT ARE RESOURCES?

Resource can be defined as "a source of supply, support, or aid." This is a simple definition for a term that is familiar and frequently used, particularly in the phrase "natural resource." But reflection should remind us that resources are identified in terms of human needs: a resource is something we want, need, or require and to which we can ascribe a real, apparent, or potential value. These measures of desirability can change rapidly with time, particularly in response to technological developments whereby new techniques create demands for materials that previously had little or no value. Uranium ore was of scant importance until the development of nuclear technology. Another example, valid at least in the United States, is the

Richard H. Fleming is professor of oceanography at the University of Washington. This paper is Contribution No. 681, Department of Oceanography, University of Washington.

3

apparent change in the value of coal deposits as other cheaper or more convenient sources of energy are exploited.

The value of a resource also depends upon its location and state or condition. Obviously, the vast outflow of the Amazon River can do little to meet the water shortage in the desert areas of North Africa. The Columbia River, in terms of existing economics, cannot be used to satisfy the ever-growing demands for fresh water in the southwestern part of our nation. Petroleum or mineral deposits in some remote or inaccessible region can hardly be considered as a valuable resource unless we can develop techniques for recovery, extraction, and transportation so that the supply can be delivered to users at competitive costs. But economics, it must be remembered, is not the sole criterion by which decisions for resource development are made. For reasons of national security, national prestige, humanitarianism, or for other social or aesthetic reasons, it may be deemed advantageous to invest large sums in the development of resources that at the moment are not competitive with existing supplies.

In any consideration of ocean resources it is essential to keep in mind that this environment, like the land and the atmosphere, is subject to various demands and uses which give rise to competition and conflicts of interest. Pollution, as a consequence of improper waste disposal, can render the environment unfit or undesirable for other purposes, creating hazards to public health as well as a reduction in desirability for recreation and enjoyment. Damming rivers for hydroelectric power plants, or for the diversion of water for agriculture and industry, may seriously impair their use as spawning grounds for anadromous fishes such as salmon. It is difficult to see how one can integrate the total value of resources or to decide upon the best course of action when there are conflicts of interest. However, these problems cannot be ignored just because they are complex and hard to solve, or because they are unpleasant to think about.

In considerations of natural resources we tend to focus our attention on items that are costly, in short supply, or apparently inadequate to meet foreseeable needs. In areas where fresh water is abundant, this essential resource is accepted as commonplace and receives scant attention from the public. However, in less fortunate areas where the supplies of fresh water are limited and the demands are approaching and even exceeding the rates of renewal, there is great concern. Many other examples could be cited to emphasize the need for a more objective attitude toward resources. We must attempt to examine them on large scales both in space and in time if we are ever to produce effective long-range plans. In many cases we must give increased attention to potential resources, those not currently being exploited because of inaccessibility, lack of technological capabilities, or comparatively high costs of development.

These considerations become particularly pertinent when we turn our attention to the oceans. Because we are land animals, to us the ocean is an unfriendly and foreign environment. Certainly the ocean has played an important part in human history, primarily for transportation and to a limited extent as a source of food. However, in broad terms it has never been exploited, and consequently our familiarity and experience with the land has not been extended to the ocean.

We must be imaginative in attempting to visualize the future importance of the ocean to mankind. In our brief history as a nation we have numerous examples of doubts and reservations expressed when we acquired by conquest, claim, or purchase the new and largely unknown lands involved in the expansion into the south and west and in the purchase of Alaska. Likewise, some twenty-five years ago, in 1946, the United States claimed possession of the resources on and beneath the shallow waters of the continental shelf adjacent to our coasts. Largely unknown and unexploited, except by mariners and fishermen, this new possession is already yielding petroleum resources, and other potential resources are being explored.

It is now trite to refer to the oceans as the "Last Frontier"; but the term is entirely appropriate because it carries with it the connotations of the unknown, the undeveloped, and the unclaimed. There is also the romance and challenge of the mysterious and remote, the promise of apparent treasure and quick wealth awaiting the bold and venturesome pioneer. Just as the quest for gold and furs attracted the pioneers into our western states, the search for the wealth of sunken wrecks and for diamonds and gold is tempting to their seagoing counterparts.

History reveals that the true wealth of our country is in the more mundane resources of our lands and waters and in our ability, through industry and an ever-expanding technology, to develop them and convert them to our purposes. The capacity of the land to produce large amounts of food and other organic materials, the deeply buried sediments to yield their hydrocarbons, the forests their cellulose, and the rocks and sediments their metals and materials for industry and construction is tremendous; but in many areas of the world large segments of the population are in dire want. It is natural that we should look to the oceans for resources in two general categories: first, those that during past centuries have been developed to some degree, such as transportation and fishing; second, those items that are scarce or in short supply on land and that consequently have a high unit value. Recovery of limited amounts of a resource or discovery of a single area of high concentration therefore promise quick returns and large profits. These hopes and aspirations may serve as desirable incentives and create the widespread interest that is essential; but, just as on land, many of the pioneers may be disappointed in their quest for individual wealth. The ultimate value of the oceans cannot be judged from our perspective. In the centuries and millennia to come, attitudes, values, and technologies will develop in ways that are as yet unknown. In our evaluation we must not be misled by past experience and current standards. Instead, we must be prepared to view the ocean with eyes sharpened by imagination and ingenuity.

We tend to give most of our attention to resources that are available in small amounts and have high unit values or to those more abundant items where unit values must be kept low, such as food and fresh water. Overproduction of resources can lead to financial crises. Violent storms, floods, and epidemics are rarely the direct consequence of human activities, but represent extreme departures from the more normal natural regimes. However, the consequences of catastrophic events can be avoided or minimized by methods of forecasting that provide adequate warning. These extreme situations emphasize the importance of being able to forecast environmental conditions that are subject to large variations. Techniques for this purpose are needed to enable us to predict not only the natural variability of environmental features but also to estimate the consequences of human activities.

Beyond the need for predictive procedures is the desirability of developing techniques of modification and control. We have had conspicuous success in modifying the land surface (sometimes to its detriment), but so far the modification of the atmosphere and of water has been largely limited to pollution—that is, undesirable changes—which is generally of only local importance. (A notable exception is the introduction of radioactive materials from the tests of nuclear devices.) Only by better understanding the processes affecting and controlling environmental conditions can we hope to develop adequate methods of prediction, modification, and control. This requirement emphasizes the fact that we are dealing with dynamic systems and also that much remains to be accomplished in the basic scientific study of the earth's environments before we can provide these methods.

THE EFFECTIVE USE OF THE SEA

In June 1966 the Panel on Oceanography of the President's Science Advisory Committee issued a report entitled "The Effective Use of the Sea." In their preamble the panel stated that a prime national objective should be for "effective use of the sea by man for all purposes currently considered for the terrestrial environment: commerce, industry, recreation and settlement; as well as for knowledge and understanding." The panel also stressed the importance of developing and improving our capabilities to predict variable phenomena and ultimately to modify and control such features of the marine environment. Furthermore, the importance of the oceans in national security and international affairs was emphasized.

This broad statement of policy is exciting and challenging. It widens the scope of our concern and provides the format for a more rational consideration of the existing and potential utilization of the ocean. It should be noted that the word "resources" does not appear, but that the emphasis is upon the *uses* of the environment by man, which obviously include the extraction of certain materials along with nonextractive activities. Attempts to compile a list of the uses of the land lead to a dozen or more major categories. There are numerous ways in which such analyses can be made, and the purposes of the list may frequently determine the most appropriate manner in which to group human needs and activities. It is generally impossible to establish clear-cut, mutually exclusive groups because of the integrated and interdependent nature of our culture. It is, therefore, more important to consider the whole rather than the individual parts. The following major categories of human uses of the terrestrial environment identify the corresponding uses of the ocean: transportation; source of food and other organic materials; source of inorganic materials; source of fresh water; source of energy; habitation; industry and business activities; transmission of energy and communications; waste disposal; health, welfare, and recreation; education and training; gaining of knowledge and understanding; national security and international relations.

Depending upon their interests and activities, individuals will tend to identify themselves with one or more of these categories and argue that these are the most important or critical to human welfare. In dealing with matters concerning natural resources and public policy, the broad, all-inclusive point of view must be developed. The apparent importance of the individual categories of uses has changed with time, largely because of technological developments. Such evaluations depend on geography and upon the level of culture in various regions: what was important a century ago may have much less significance today, while other uses are rapidly rising in importance.

The most casual examination of the existing and potential uses of the ocean in these categories reveals that there are certain of them that have traditionally supplemented or competed with the land. For thousands of years water in general, and sea water in particular, has been used for transportation and as a source of food, salt, and minerals. At the other extreme, the ocean has never been exploited for habitation or for the conduct of business and industry—activities that can be conducted more conveniently and comfortably on land. In many of the other categories we can identify developments and changes that have occurred in recent decades, mainly because of increased technological capabilities, that have created new demands for materials or uses or that have made possible new enterprises. The recovery of petroleum from beneath the sea bed is an example of the latter.

It is not the purpose of this review to describe the state of development of marine activities included in the various categories. What is important is to realize that there are innumerable possibilities for new and innovative developments. Past history is a poor guide to future potentials. Our needs are changing so rapidly and our technological capabilities developing at such rates that we must continually revise and update our estimates of what can and cannot be done, of what is desirable and what is less appropriate, and of what priorities should be assigned to various enterprises.

Scientists and engineers have the responsibility of providing guidance in terms of technical possibility and feasibility, but many others will be involved in decision-making processes. It is essential that ocean-oriented lawyers, economists, business experts, sociologists, and others be available and that legislative bodies and the general public become more familiar with the ocean. Many matters of public policy must be determined, long-range plans must be devised and properly implemented. The problems are complex, experience is often inadequate, and the best of cooperative efforts will be required to arrive at wise decisions.

THE PLANNING PROCESS

To decide upon the desirability of undertaking a program for the use of the ocean, it is necessary to proceed through an orderly sequence of analyses. In this process there are four major steps: (1) *identification* of the contemplated resource or use; (2) *evaluation* of the technological, legal, economic, sociological, and political factors; (3) *estimation* of the benefits anticipated, costs involved, goals, needs, and time schedules; and (4) *implementation* through arrangements for funds, authorization, and assignment of responsibilities for management and regulation.

The initial stage, identification, is largely the task of scientists and engineers. It is their responsibility to provide and/or compile information pertinent to the resources or use under consideration. Where or what is it? What conditions may affect concentration, forms, and total abundance of resources? Where are they located? In part, this information may be available from programs of systematic surveys carried out by government or industry, but it is the general experience that these sources are inadequate in detail or accuracy and that additional surveys and studies will be required.

Evaluation is the process of integrating a wide variety of data bearing on the problem. Information dealing specifically with the resource or use must be judged in terms of the influence that the environment may have upon contemplated operations. What influence will conditions on, over, in, or beneath the sea have upon the activities and what may make them too dangerous or too costly to be practicable? The ocean is in general hostile to the materials and techniques commonly employed on land; activities that can be accomplished with ease on land may prove very difficult to execute in the sea. Economic advantages must be considered with respect to competitive practices on land and abroad and with regard to future trends. The political and legal aspects of the proposed operations must be reviewed. Recent decades have seen the traditional concept of the freedom of the seas giving way to national claims to the resources of the coastal areas, and to treaties and international conventions concerning certain areas, resources, and uses. These too are undergoing rapid changes as national interests and technological capabilities alter the real or apparent value of the ocean.

Estimation requires that goals be defined in terms of objectives, levels of anticipated activity, and a schedule of achievement. From the information compiled and understanding gained during the processes of identification and evaluation it should be possible to establish rational goals and to outline realistic programs to achieve them. Because of the nature of the ocean and the lack of industrial initiative, much of the long-range planning has been done by government agencies or by scientific groups encouraged or sponsored by the federal administration. Recently various international bodies, including the United Nations, have shown increased interest in this area. These developments are a natural consequence of the character of the ocean and the realization that most of its features are of international and national concern. The characteristics of the ocean do not conform to national, state, or local political boundaries, and the practices of localized regulations and management followed on land may interfere with and retard maritime developments that should be conducted on much broader regional or even ocean-wide scales.

National plans and programs prepared by various groups have, in general, not proved to be very satisfactory. Those prepared by scientists tend to emphasize the importance of basic research to the detriment of other types of activities. Federal programs endorse existing agencies and foster their expansion to the exclusion of new and innovative projects that should be encouraged. Industry, seeking quick returns, promotes short-range goals with major support from the government. In 1967, two groups created by federal legislation under the Marine Resources and Engineering Development Act of 1966 began comprehensive, long-range studies to establish goals and national programs for the oceans. The National Council on Marine Resources and Engineering Development was composed of high-level administrators from a number of federal agencies. The National Commission on Marine Science, Engineering and Resources was made up of representatives drawn from the academic world, industry, and business. It was the responsibility of this commission to define long-range objectives and to recommend the organizations, budgets, and programs required to achieve them. These, then, were first attempts to arrive at truly comprehensive programs with long-range goals. The passage of the National Sea Grant College and Program Act of 1966 is further evidence of the national concern for fuller utilization of the ocean. At the state level this same interest was demonstrated by the creation of the Oceanographic Commission of Washington by legislative action in 1967. The University of Washington responded to these broader concepts by establishing in 1967 a Division of Marine Resources to coordinate and support the activities of the many academic units engaged in marine sciences and engineering and to foster a more complete utilization of the ocean.

Implementation of programs in terms of goals, levels of activities, and time scales may involve legislative action, the issue of leases and permits, and negotiations of many kinds. Decisions as to whether a program should be undertaken, accelerated, or retarded will depend upon the desirability of the program itself; but in virtually all cases the situation will be affected by external factors. Prestige, humanitarian considerations, and standing agreements may also enter the decision-making process.

RECOMMENDATIONS

The ocean, covering nearly three-quarters of the earth's surface, affects human activities both directly and indirectly. It possesses vast potential to meet human needs. In a period of rapid technological developments, the capabilities to utilize the ocean and to exploit its resources are undergoing great changes. Existing practices are generally obsolete and outmoded. The problems of the oceans are clearly national and international, involving large regions rather than small areas. This complex situation must be recognized and steps taken to ensure that far-sighted plans are developed and programs undertaken to achieve clearly defined goals and objectives. To avoid the undesirable consequences of hasty decisions made on the basis of inadequate understanding, and to proceed in an effective way, it is imperative that we improve our communication, cooperation, and coordination.

Communication requires a fuller and more rapid exchange of information and understanding among a wide spectrum of agencies and special groups. Cooperation is essential if duplication and expensive mistakes are to be avoided. Better means for full cooperation between government, industry, and the universities must be devised so that each may contribute most effectively to the development of the ocean. Agreement and coordination must be realized in establishing goals, devising plans, and in conducting programs so that balanced progress may be made in all aspects of maritime activities. Missions, roles, and responsibilities may require redefinition.

Much must be accomplished in the coming years in the development of the uses of the ocean. Many questions affecting public policy must be analyzed and answered. The challenges are great, but the rewards will be of tremendous significance to the generations to come. Regardless of background, experience, and responsibilities, everyone is involved and must be prepared to identify and solve these matters relating to ocean resources and public policy.

Coastlines and Continental Shelves: Geological History and Characteristics

JOE S. CREAGER

COASTLINES and continental shelves as we recognize them today are, in a geologic sense, transient physiographic features. The morphology of the present continental shelf is primarily the result of fluctuations of sea level during the Plio-Pleistocene. The present coastlines are the physiographic features resulting from the actions of subaerial and marine erosional and depositional processes along a shore of relatively constant position over a span of time, the length of which is dependent upon local rates of sea level and landmass rise and fall.

Coastlines and continental shelves are only subfeatures of the extensive continental terrace. A hypsographic curve (Fig. 1) of the earth's surface reveals that the relief of the surface lies on two dominant levels: one, within a few hundred meters of sea level, represents the normal surface of the continental blocks; the other, between 4,000 and 5,000 meters below sea level, represents the ocean-basin floor. The ocean basins and the continental blocks are distinctly different in topography, structure, lithologic types, and, in all probability, age. The continental slope (Fig. 2) is the noticeable change in elevation between these levels that marks the structural edge of the continent and overlies the change from thick continental crust to thin oceanic crust. A broad look at some of the details of a portion of this transition zone is the subject of this paper.

The major portion of the waters of the earth's surface occupies the large structural basins. Oceanic waters presently overfill these basins, shallowly submerging the outer margins of the continental surfaces. The shallow submerged platform bordering the continents is the continental shelf (Fig. 2). It slopes seaward with a world-wide average gradient of approximately $0.1°$. The seaward edge is marked by an increase of gradient to an average of $4°$

Joe S. Creager is professor of oceanography at the University of Washington. This paper is Contribution No. 682, Department of Oceanography, University of Washington.

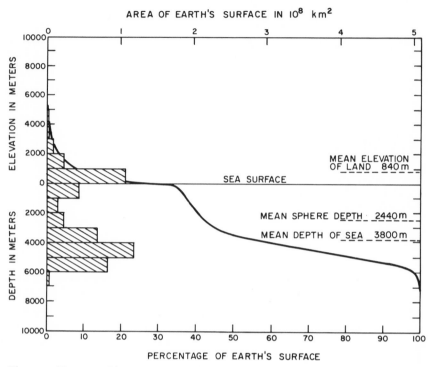

Figure 1. Hypsographic curve showing the area of the earth's solid surface above any given level of elevation or depth (after Sverdrup et al. 1942, p. 19; by permission of Prentice-Hall, Inc., Englewood Cliffs, New Jersey)

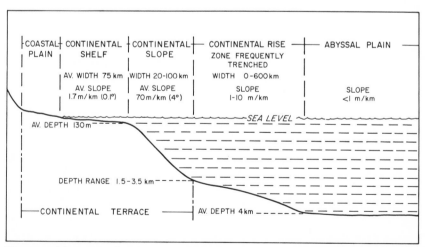

Figure 2. Diagrammatic profile of continental terrace with continental rise, showing average or range of depth, width, and slope of component parts (after Curray 1965; by permission of Princeton University Press)

down the continental slope. Although the average depth of this line of change in gradient, or shelf break, averages about 130 meters, there is considerable variation from a few tens of meters to several hundred meters. The term "shelf" is commonly restricted to those features that terminate at a depth of less than 550 meters (Shepard 1963, p. 206); however, by the United Nations Convention on the Continental Shelf (1958), the shelf is terminated at 200 meters or undefined exploitable depths. Although this seminar is designed to consider specifically the ocean's resources and therefore makes the legal definition pertinent, this paper will use only the definition based upon a change in gradient of the bottom along the shelf break. The base of the continental slope may terminate in gently sloping deposits of sediments that variously overlap the base of the continental slope and extend seaward—in some cases hundreds of kilometers over the ocean floor. These deposits are the continental rises. Where they are not present, the base of the continental slope frequently terminates in an oceanic or deep-sea trench.

The continental terrace is defined as the rock and sediment mass beneath the continental slope, continental shelf, and coastal plain. The upper surface of the continental terrace does not terminate landward at the coastline, but rather is a single geologic province termed coastal plain to landward and continental shelf to seaward. The coastline or dividing feature is continually shifting and undergoes wide changes in position depending upon both local and eustatic changes in sea level and upon local rates of erosion and deposition.

Today we see the coastline as the inner margin of the flooding of the continental surface by the ocean. The character of the coastline in any area is a function of (1) the nature of the solid earth mass, that is, its lithology, structure, topographic character, and vegetation cover; (2) the interplay of processes active along the coastline resulting from glaciation, earth movements, runoff, winds, waves, currents, chemical action, and the activity of organisms; and (3) the time during which the processes have acted. Without becoming enmeshed in the myriad possibilities of coastline types and origins, it is sufficient to inspect briefly two extreme types, a coastline of cliffs and marked relief, and a coastline associated with a broad, low-relief coastal plain.

The cliffed coastline normally develops along a coast closely bordered by young mountains providing high relief (Fig. 3A). Drainage basins in such areas are usually small and runoff is lower than average, supplying the coastal region with relatively small amounts of sediment. The form the coastline takes is then controlled by the composition of the mountain masses, their relief, and the action of marine processes, primarily wave action. In higher latitudes glaciation may be important and in lower latitudes colonial organisms may control the morphology of the coastline. Erosion by

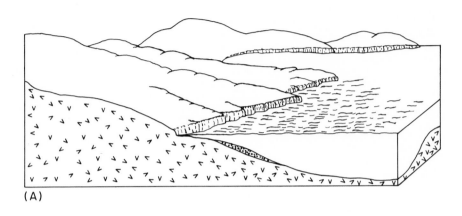

(A)

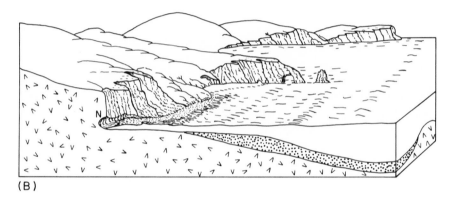

(B)

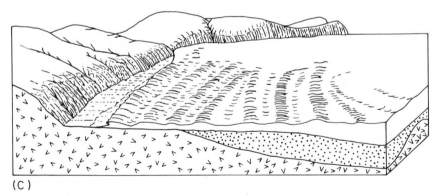

(C)

Figure 3. Development of a cliffed coast (after Strahler 1960; by permission of John Wiley & Sons, Inc.)

wave action is most effective on the rock masses with least mechanical resistance. Those portions of the coast comprised of such easily erodible rock retreat landward more rapidly than other portions, producing a highly indented coast, whether or not this was the case initially (Fig. 3B). The resulting irregular bathymetry refracts the incoming waves, focusing them on the headlands and increasing relatively the erosion of these promontories. Erosional products from the headlands are swept into intervening bays by longshore currents directed away from the headlands. The longshore currents are the result of incomplete wave refraction.

Assuming sufficient time is available for this particular geomorphic cycle to go to completion, the end result, as seen in Figure 3C, is a relatively straight coastline set back some distance from the original. The cliffs are higher. The sea bottom, or inner continental shelf, off the cliffs is a gently sloping eroded rock platform that may be extended seaward beyond the original coastline as a depositional terrace conformable with the eroded terrace. With further time the cliffs will be reduced to lower slopes by weathering and runoff, and the eroded platform may become shallowly buried by a veneer of sediment.

The coastline associated with a broad, low-relief coastal plain is quite different in character. This coastline is primarily dependent upon the rate of sediment supply from the land and the rate at which marine processes are able to redistribute the sediment. Drainage basins in such areas are usually larger than average, providing the coastal regions with relatively large amounts of sediment. The bulk of the sediment is frequently supplied by a single river system as essentially a point source along the coast. The remainder of the coast receives progressively smaller quantities of sediment in both directions away from this point source. If sea level has not recently risen to produce embayed estuaries at river mouths or if the river has supplied sediment in such quantities as to prevent the formation of an estuary, a delta is constructed at the coastline (Fig. 4). The shape and seaward extent of the delta is a function of river sediment load, the continuity of sediment discharge throughout a year, and the ability of waves and currents to redistribute the sediment deposited in the delta. If the sediment discharge is large and continuous throughout the year, the delta is large, sprawling, and irregular, such as the Mississippi River delta. Where the sediment discharge is less and confined to distinct flood periods, the delta is smaller and more regular in plan shape, such as the Rio Grande delta. With time, the river migrates laterally across the width of its flood plain, building its delta at various positions.

The areas to either side of a delta receive sediment from the delta region by wave and current action and from local small rivers and streams; or, if these sources are not within range, small quantities of sediment may be supplied from shore and nearshore erosion by waves. In locations where along-

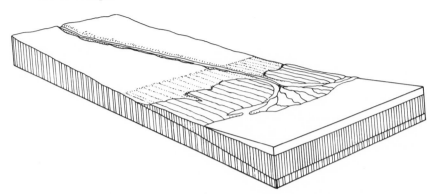

Figure 4. Delta coast (after Strahler 1960; by permission of John Wiley & Sons, Inc.)

shore supply of sediment is limited, a cycle of shoreline development, shown in Figure 5, may prevail. Sediment transported shoreward along the bottom by wave action is deposited along the line of breaking waves to form an offshore bar (Fig. 5A). As this bar is built upward, a chain of low-lying islands is produced (Fig. 5B), and with time these islands become a more continuous belt of land, termed a barrier island (Fig. 5C), separated from the mainland by a lagoon. The lagoon is gradually filled by sediment supplied from landward, from seaward through tidal inlets, and across the top of the barrier islands during storm wave conditions. A tidal marsh then takes the place of the former lagoon (Fig. 5D). With limited supply of sediment and transfer of sediment from the seaward face of the barrier to the lagoon face, the barrier migrates landward, destroying the lagoon and marsh. The near-shore bottom is now deeper than before, permitting larger waves to reach close to the original shoreline before breaking.

A similar process of offshore bar formation may evolve nearer to a major source of sediment—that is, nearer the delta in our example—with additional bars being formed and built above water before enough time has passed for the original bar to be driven ashore and destroyed. A strand plain (Fig. 6) composed of abandoned bars and intervening low marshy areas totaling 250 distinct ridges in places has been discussed by Curray and Moore (1964). Such a coast is a prograding one, as are deltas, in that the coastline is being built seaward.

Only if sea level remains at a relatively constant position for some time do the cliffed or low coastal plain coastlines develop through the cycles discussed. It is generally accepted that through geologic time, sea level has not remained statically at a fixed elevation, but has risen and fallen approximately 100 meters numerous times (see, for instance, a rather complete discussion in Kuenen 1950, pp. 532–41). More recently, during the periods of extensive continental glaciation, sea level has fluctuated by as much as 140 to 160 meters below its present level (Donn et al. 1962), and it is estimated

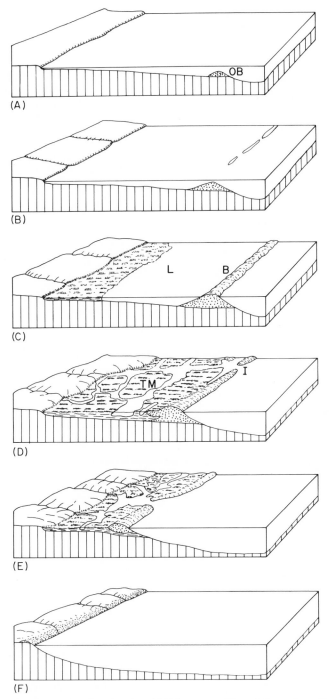

Figure 5. Development of a barrier island coast; OB = offshore bar, L = lagoon, B = barrier island, I = inlet, TM = tidal marsh (after Strahler 1960; by permission of John Wiley & Sons, Inc.)

Figure 6. Oblique air view of abandoned bars and intervening low marshy areas. Average width of ridges shown is about 80 meters (photograph by F. B. Phleger, reprinted from Curray and Moore 1964 by permission of the authors and the American Association of Petroleum Geologists)

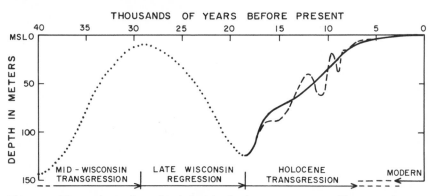

Figure 7. Fluctuations of sea level during the last 40,000 years (Curray 1965; reprinted by permission of Princeton University Press)

that a complete melting of ice on the earth's surface today would produce a eustatic rise of sea level of about 80 to 100 meters. At least four major glaciations during the Pleistocene epoch are generally accepted and have caused sea level changes in the ranges suggested above. The most recent significant shift in sea level produced a maximum lowering of approximately 105 to 120 meters below present level during the late Wisconsin (18,000 years Before Present) glacial advance (Fig. 7). This lowering was followed by a rapid, but fluctuating, rise (Holocene transgression), with the rate of rise markedly decreased about 7,000 years B.P. During the last few thousand years, sea level has been near its present position, with some possible minor fluctuations above and below (see Curray 1965 for a more complete discussion).

At the time of the most recent glacial lowering of sea level, and also possibly during previous glacial lowerings, the coastline was near the shelf break. A stillstand of sea level at a position near the shelf break resulted in the formation of coastline features at this position and the deposition of all runoff-distributed sediments directly on the upper portion of the continental slope. Exposed landward was a broad coastal plain, which is now, in part, called the continental shelf. This plain was subject to all of the subaerial processes of erosion and deposition recognized today on our present coastal plains. Rivers were rejuvenated with increased vertical fall from their headwaters and incised valleys across the plain. Both the rivers and runoff from the interfluves eroded sediment present on the plain, producing an erosional surface whose character depended upon the amount of sediment cover, resistance of the rock material beneath the sediment cover, and the magnitude of the drainage basin.

As sea level rose, the coastline shifted landward and the zone of processes active in the formation of a coastline migrated landward. During the periods of most rapid sea level rise, recognizable coastline features did not develop. However, at the positions of extremes in the minor fluctuations, sufficient time was apparently available to permit the development of recognizable features, which have been used to generate the sea level curve presented in Figure 7. As sea level rose, a veneer of sands, associated with the shallow nearshore zone of active wave-controlled erosion, distribution, and deposition, was deposited over the subaerially produced coastal plain surface. The thickness and extent of such a deposit depends largely upon the sediment supply during transgression. In regions of high relief where cliffed coasts developed, such as on the west coast of the United States, the supply of sediment was low and the transgressive sands were thin, whereas off broad coastal plains where the supply is greater, as on the Gulf and Atlantic coasts of the United States, the transgressive sands were comparably thicker. During minor sea level stillstands, there are discontinuities in this transgressive sand veneer which are recognized by increased sediment thickness. In some cases, this discontinuity is seen in greater relief of the subsurface which is caused by wave erosion and then complete or partial burial. Figure 8 is a subbottom acoustic record of a feature off Point Loma, California, which has been interpreted as a partially buried cliffed coastline. Figure 9 is a bathymetric map of Hope Valley, which crosses the continental shelf of the Chukchi Sea. It has been concluded that a deltaic valley fill at a coast

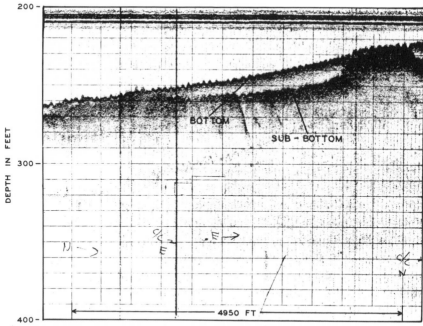

Figure 8. Subbottom acoustic record of a feature off Point Loma, California, interpreted as showing a partially buried old cliffed shoreline (Moore 1957)

during one of the minor sea level stillstands has rendered a portion of the valley not distinguishable in the present bottom topography (Creager and McManus 1965).

The nearshore zone where wave action is important as a geologic event in the formation of coastal features rarely extends seaward to depths greater than ten meters (Dietz and Menard 1951; Dietz 1963; Moore and Curray 1964; Dietz 1964). Seaward of this depth, the finer sediments carried in suspension are deposited (see Curray 1965 for a more complete discussion). This sediment is absent, locally present, or widely distributed over the continental shelf surface seaward of a depth of ten meters, depending upon the sediment supply and the current and wave systems present. The surface of a continental shelf may be exposed bedrock or variously covered laterally and at depth by nearshore or open continental shelf sediments. The sediments, or lack of sediments, and the surface configuration of the continental shelves as known today are a product of erosion and deposition by both marine and subaerial processes during sea level fluctuations of the Plio-Pleistocene. Differences among continental shelves may be attributed to local differences in process.

For convenience, the shelf sediments have been grouped into the following five types (Emery 1952): (1) *authigenic,* consisting of minerals such as

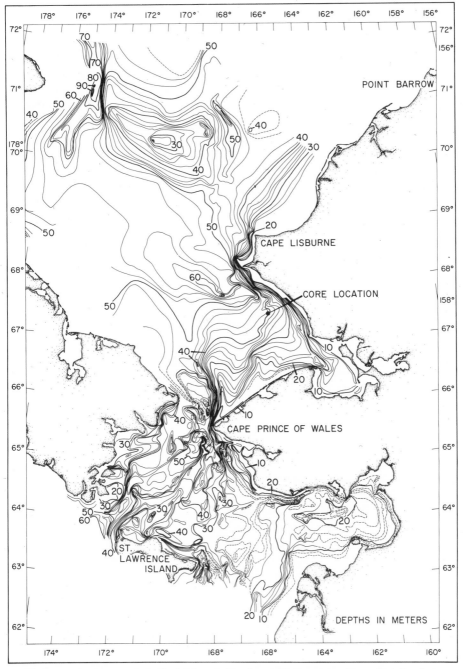

Figure 9. Bathymetry of Chukchi and Northeast Bering seas. Note that the long northwest-southeast shelf valley is not distinguishable in the contours at the core location because of a deltaic valley fill at an old coast during a minor sea level stillstand (adapted, with permission of the publishers, from Joe S. Creager and Dean A. McManus, "Geology of the Floor of Bering and Chukchi Seas—American Studies," in David M. Hopkins, ed., *The Bering Land Bridge* [Stanford, Calif.: Stanford University Press, 1967])

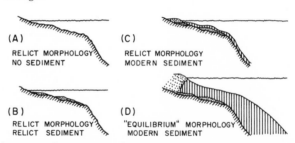

Figure 10. Diagrammatic summary of possible relict (inherited from previous environmental conditions) versus equilibrium sediment and shelf morphology combinations (Curray 1965; reprinted by permission of Princeton University Press)

glauconite and phosphorite which formed *in situ;* (2) *organic,* which includes foraminifera and other shells, the skeletal matter of organisms; (3) *residual,* having weathered from underlying rocks; (4) *relict,* or remnant from a different earlier environment; and (5) *detrital,* presently being supplied by rivers, currents, or wind. It is important to consider also whether the sediment present at any location is modern, in the sense of being in equilibrium with present environmental conditions of the depositional area, or relict, in not being in equilibrium with present conditions. This difference is apparent and implicit in considering relict or detrital sediment by the proposed categorization, but authigenic, organic, or residual sediments can be either relict or modern. Concomitant with this consideration is the importance of distinguishing between relict and modern surface morphology, depending upon whether or not it is in equilibrium with its present environment. Curray (1965) has summarized the possible combinations and distribution of relict versus modern sediments and morphology of the shelf surface (Fig. 10).

An understanding of the details of the geological history and characteristics of coastlines and continental shelves is of utmost importance to the problems of location and recovery of economically important mineral deposits on the continental shelves. Minerals have been concentrated on the continental shelves by the same physical and chemical processes that have led to the present shelf morphology and sediment distribution. Possibly the most important resources in terms of overall value are the sand and gravel deposits. Extensive areas are covered by relict transgressive sands. The coarsest, and thus most useful, materials occur beyond the present active zone of coastline formation, making it possible to mine these deposits without affecting the nearshore processes. These same relict sands and gravels are the most likely locations of important concentrations of authigenic minerals, for the fact that they are relict implies that sufficient time has been available for the accumulation of significant minerals requiring extremely slow depositional rates of diluting materials. Two such authigenic minerals have known economic potential: one of these is phosphorite, which is a phosphate fertilizer; the other is glauconite, a hydrated potassium, iron, aluminum silicate. It has been used as a soil conditioner and water softener and might be important as a source for potash.

Placer deposits of both beach and stream types exist on the continental shelf. In some parts of the world, both have been mined successfully for tin, magnetite, ilmenite, chromite, gold, and diamonds. Placer deposits are the lag concentrations of heavier minerals produced by the action of moving water either in the form of the oscillating motions of water in waves and associated longshore currents or unidirectional flow in rivers and streams. Both river and beach placers can be expected on the now submerged continental shelf.

This brief report was not prepared as a summary of our knowledge of mineral deposits present on continental shelves, but as a summary of the pertinent parts of the history and characteristics of the shelves that would be useful to someone interested in further investigation of the mineral resources present on this portion of the earth's surface. Mero (1965) has summarized in some detail our knowledge of the mineral resources on the continental shelves.

REFERENCES

Creager, Joe S., and D. A. McManus
   1965   "Pleistocene Drainage Patterns on the Floor of the Chukchi Sea." *Marine Geology* 3:279–90.
Curray, Joseph R.
   1965   "Late Quaternary History, Continental Shelves of the United States." Pp. 723–35 in H. E. Wright, Jr., and David G. Frey, eds., *The Quaternary of the United States*. Princeton, N.J.: Princeton University Press.
———, and David G. Moore
   1964   "Pleistocene Deltaic Progradation of Continental Terrace, Costa de Nayarit, Mexico." American Association of Petroleum Geologists Special Publication, *Marine Geology of the Gulf of California*, Memoir No. 3, pp. 193–215.
Dietz, Robert S.
   1963   "Wave-base, Marine Profile of Equilibrium, and Wave-built Terraces: A Critical Appraisal." *Geological Society of America Bulletin* 74:971–90.
   1964   "Wave-base, Marine Profile of Equilibrium, and Wave-built Terraces: Reply." *Geological Society of America Bulletin* 75:1275–82.

Dietz, Robert S., and H. W. Menard

1951   "Origin of the Abrupt Changes in Slope of Continental Shelf Margin."
Bulletin of the American Association of Petroleum Geologists 35:1994–
2016.

Donn, William L., W. R. Farrand, and M. Ewing

1962   "Pleistocene Ice Volumes and Sea-Level Lowering." *Journal of Geology* 70:206–14.

Emery, K. O.

1952   "Continental Shelf Sediments of Southern California." *Bulletin of the
Geological Society of America* 63:1105–8.

Kuenen, Philip H.

1950   *Marine Geology.* New York: John Wiley & Sons.

Mero, John L.

1965   *The Mineral Resources of the Sea.* New York: Elsevier.

Moore, D. G.

1957   "Acoustic Sounding of Quaternary Marine Sediments off Point Loma,
California." *Navy Electronics Laboratory Report,* no. 815.

———, and J. R. Curray

1964   "Wave-base, Marine Profile of Equilibrium, and Wave-built Terraces:
Discussion." *Geological Society of America Bulletin* 75:1267–74.

Shepard, Francis P.

1963   *Submarine Geology.* 2nd ed. New York: Harper & Row.

Strahler, Arthur N.

1960   *Physical Geography.* 2nd ed. New York: John Wiley & Sons.

Sverdrup, H. U., M. W. Johnson, and R. H. Fleming

1942   *The Oceans: Their Physics, Chemistry, and General Biology.* New
York: Prentice-Hall. Copyright renewed, 1970.

United Nations

1958   United Nations Convention on the Continental Shelf, 29 April [effec-
tive 1964]. 15 U.S.T. 471, T.I.A.S. No. 5578, 499 U.S.T.S. 311 (U.N.
Doc. No. A/CONF. 13/L.55).

Physical Oceanography of Estuaries

MAURICE RATTRAY, JR.

AN estuary is a region of transition from river to ocean. Its definition will vary from author to author, but probably the most useful definition for our purposes is that given by Pritchard (in Lauff 1967): "An estuary is a semi-enclosed coastal body of water which has a free connection with the open sea and within which sea water is measurably diluted with fresh water derived from land drainage." Another common definition includes that portion of a river in which the tidal effects are felt; but for our purposes it is more convenient to call that portion of a river extending above the region of appreciable salinity intrusion to the upstream limit of tidal effects a tidal river, and to retain the term estuary for those portions in which the salinity is increased by salt supplied from the ocean.

Three kinds of estuaries can be identified by their geomorphology (Cameron and Pritchard 1963). Coastal plain estuaries are essentially wide rivers, flowing across coastal plains into the sea. They are usually shallow and may have a dredged channel for navigational purposes. Examples are Chesapeake Bay on the East Coast, the Mississippi River on the Gulf Coast, and the Columbia River on the West Coast. The fjord is a deep, glacially scoured inlet, which in many cases has a sill at its mouth. Examples are Hood Canal, Silver Bay, Alaska, and the many inlets in British Columbia and southeast Alaska, as well as the fjords in Norway. And finally there are bar-built estuaries in which bays or lagoons are separated from the open sea by bars through which there may be one or more channels. These occur commonly on the Gulf Coast. In many ways they behave more like a small sea or a lake than like either a coastal plain or fjord estuary; that is, they will have an important lateral circulation and relatively minor effects be-

Maurice Rattray, Jr., is chairman of the Department of Oceanography at the University of Washington. This paper is Contribution No. 683, Department of Oceanography, University of Washington.

25

cause of the varying density of the waters. For our purposes, we will consider only the behavior of the coastal plain and fjord estuaries.

The prime factors governing an estuary's distribution of properties and currents, other than its shape, are the tide, the river inflow, the salinity (and thus density) difference between the fresh water and the seawater, and the wind. The tide in an estuary enters from the ocean; as the tide wave progresses up the estuary, it gradually loses its energy through frictional processes and, where there are changes in cross-sectional area, is partially reflected back toward the open sea. Those tides in which a relatively small percentage of the incoming tidal energy is reflected back to sea and in which, therefore, most of the tidal energy is dissipated within the estuary, are called progressive wave tides. In such cases, the times of high and low water are determined by the rate of progression of the shallow water tide wave in from the open ocean. For example, this applies very well to the tide progressing up the Columbia River. At the other extreme is the standing wave tide, where the major portion of the incoming tide energy is reflected back out to sea with relatively little dissipation in the estuary itself. In this case the time of high and low water is simultaneous, or almost simultaneous, throughout the estuary. With standing wave tides, there is the possibility of resonance, with an increase in tidal elevation going from the mouth to the head of the estuary; examples are the high tides found in the Bay of Fundy and in Alaska's Cook Inlet. The Puget Sound system has a tidal regime intermediate between the two extremes. The tidal elevation does increase toward the head of Puget Sound; the time of tide is not, however, quite simultaneous, as there is some delay of the tide in the lower sound over that found at the entrance.

Vertical tidal excursions are quite regular, composed of diurnal and semidiurnal components in the same manner as in the open sea. However, in long, shallow estuaries, distortion of the tidal wave is caused by nonlinear effects, with high water progressing more rapidly up the estuary than low water. Tidal currents, on the other hand, are highly irregular and consist of motions over many scales of space and time; only when appropriate averaging is used can a smooth variation of these currents be identified. Tidal currents reverse first in those regions of larger friction, such as near the shore, over shoals, or behind points. Bathymetric effects can result in local net flow, eddies, and tide rips.

With density stratification, internal tidal waves may be expected and in particular in those estuaries which have large changes in depth, such as with the separate basins and sills found in Puget Sound. Tidal currents, in addition to causing an oscillating movement of salt in and out of the estuary with the tidal period, will effect a net transport of salt due to correlation of the velocity and salinity fluctuations. At tidal scale, this appears as a net advection of salt due to tidal currents.

Tidal currents provide energy for smaller-scale turbulent motion which constitutes the prime frictional mechanism acting on the tidal current and is also effective in the turbulent diffusive transfer of salt and heat. Where currents are weak, such as in fjords, the turbulence must derive its energy from the mean motion resulting from the river runoff. In other cases where turbulence is totally suppressed, diffusive transfer must depend on molecular processes alone.

The river or rivers flowing into an estuary provide an excess of water which flows through the system and out to sea with a magnitude that will vary with the inflow and the rate of change of storage within the estuary. This water experiences an increase in salinity by diffusion so that as it flows out, it removes salt from the system. Salt balance is maintained by a return circulation from the ocean and by horizontal exchange due to tidal and turbulent mixing processes. Wind-induced circulation, either up or down the estuary at the surface, depending upon the wind direction, may modify the salt balance. Because the inflowing saline ocean water required for salt balance is more dense than the river water, it will flow at depth underneath the surface layer.

Currents and other physical properties in an estuary have many scales of variation both in time and in space. In the time domain, fluctuations associated with turbulence and seiches occur for periods less than tidal. Between the semidiurnal and diurnal tidal periods, in addition to the tides themselves, are circulations induced by diurnal wind and insolation variations. Tidal variations, runoff variations, and weather cycles contributing to fluctuations occur in daily to fortnightly periods. Periods longer than fortnightly occur on account of seasonal cycles and other long-period weather patterns.

Consideration of spatial scale is more difficult because there is no simple sorting of the scales associated with the various modes of estuarine motion. There is also an important difference between the scales of vertical and horizontal processes or distributions. Vertical scales are affected by the density stratification; greater stability is associated with smaller vertical length scales. Turbulence associated with the nondeterministic portions of the velocity field will have scales ranging from the viscous cutoff, roughly on the order of a centimeter or less, to a significant fraction of the distance to the nearest boundary, which for the vertical scale would be the depth and for the horizontal scale, the width of the estuary. Density stratification would diminish the upper limit of the turbulent vertical scale. The scales for tidal variations would be a significant fraction of the total depth for the vertical and a significant fraction of the width and/or the horizontal tidal excursion for the horizontal. The circulations associated with the density field or wind-driven circulations again have scales ranging from a small fraction of the total depth to the width or length of the estuary. Thus we see in both the horizontal and vertical that the maximum scales of all the modes of motion are comparable, but that the smaller-scale motions will be due almost entirely to the turbulence.

It is always necessary when measuring or discussing the properties of an estuary to include consideration of an appropriate averaging technique: even the most idealized example would have an essential separation between a deterministic mode and a turbulent mode which must be described statistically in terms of appropriate averages. The most direct method, both mathematically and conceptually, for averaging is to use the so-called ensemble average, in which one visualizes repeated experiments on identical systems with the same external conditions and compute the average from the results of the individual observations. This procedure has a simple statistical interpretation and avoids the difficulty arising from the overlapping of scales in space and time for the various modes. However, as a practical matter, an ensemble average is impossible to obtain in real estuaries. Separation between the turbulent and deterministic modes is usually made in the time domain, with an averaging time selected which will average the effects of the turbulence and yet retain the slower variations of the other motions. Because of overlapping scales, a similar sort of average in the spatial domain is not as useful for this particular purpose. On the other hand, further averaging in the space and time domains has long been used in order to simplify descriptions of estuarine behavior.

However, there are numerous occasions in which inappropriate averaging has given misleading results, and extreme care is needed to make only averages appropriate to the problem at hand. In many cases, because of nonlinear effects, mathematical models of large-scale distributions must include scales much smaller in order to include all the relevant processes effecting changes in the variables. A limiting consideration is that averaging should not be carried out for scales larger than those for the phenomena of interest. For example, with salinity distribution in many cases it is perfectly adequate to average the relatively small variation across the width of an estuary. On the other hand, if one is interested in a constituent from a local source, the cross-channel average concentration near the source may give no useful information whatsoever. Similarly, for many purposes one is concerned only with the value of salinity averaged over a tidal cycle. Such an averaging time, however, would be essentially useless for discussing the distribution of a chemical with a reaction time that is a small fraction of the tidal period.

In association with the choice of appropriate scales for a phenomenon, one also finds that certain terms in the governing equations may or may not be important. Again, the relative importance of the various types of balance terms may be different between variables in the same estuary at the same time, so that an appropriate model for one distribution may not, in fact, be appropriate to another. One must be very careful in selecting appropriate models for the variables of interest in an estuary.

Recent studies of the dynamics of estuaries usually have been of two kinds. The first is a study of the tidal regime in which the distributions of tidal elevations and tidal currents are determined, which then permits an investigation of the effect of these currents in dispersing any material. This procedure is useful particularly in the tidal portion of rivers above the limit of salt intrusion (see, for example, Leendertse 1970). The other is an investigation of the currents and density regime averaged over a period of one or more tidal cycles. This technique is applicable to the study of distributions which have time scales longer than a tidal cycle. In any case, the effect of the horizontal salinity gradient in estuaries is important and must be included wherever present.

The tidal dynamics of estuarine systems are reasonably well understood. The current speed is essentially independent of depth, except for frictional effects. It is adequate to consider the effect of friction only in terms of a bottom stress. Obtaining a solution of the problem, however, in many cases requires a complexity of numerical integrations demanding a computer. The complications arise not from inadequacy of understanding, but from irregularities in the geometry.

In the time-average circulation, on the other hand, we have a much more complicated dynamic system, in which interaction of the current and the density fields must be considered, but which is relatively insensitive to moderate changes in geometry. Because of their narrow, elongated shape, many estuaries have only unimportant lateral variations. Variations in properties permitting a classification scheme depend upon two parameters: one, a measure of the circulation, and the other, a measure of the stratification. Different estuarine regimes occur for different values of these parameters.

A given estuary can be placed or located on this stratification-circulation diagram from a knowledge of its bulk parameters: the river runoff, the depth of the estuary, the tidal speeds, and the density difference between the source sea water and fresh water. Figure 11 shows this classification of estuaries on a stratification-circulation diagram. Seven types of estuaries are identified. In Type 1 the net flow is seaward at all depths, and upstream salt transfer is affected by diffusion. Type 1a is archetypical of a well-mixed estuary in which salinity stratification is slight, while in Type 1b there is appreciable stratification. For Type 2 the net flow reverses at depth; both advection and diffusion contribute importantly to the upstream salt flux. The stratifications in Types 2a and 2b correspond to those for Types 1a and 1b, respectively. Type 3 is distinguished from Type 2 primarily by the dominance of advection in accounting for over 99 percent of the upstream salt transfer. In Type 3b estuaries the lower layer is so deep that in effect the salinity gradient and the circulation do not extend to the bottom, an important qualitative difference from other types of estuaries. Fjord estuaries are generally of Type 3b until mixed to the extent that they assume the Type 3a

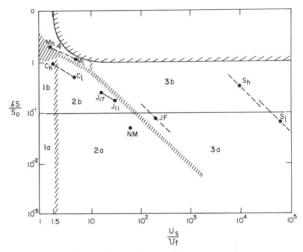

Figure 11. Proposed stratification-circulation classification, with some examples. (Station code: M, Mississippi River mouth; C, Columbia River estuary; J, James River estuary; NM, Narrows of the Mersey estuary; JF, Strait of Juan de Fuca; S, Silver Bay. δS = top to bottom salinity difference at the given section; S_0 = average salinity at the given section; U_s = net velocity at upper surface; U_f = river discharge per unit area of the given section. Subscripts h and l refer to high and low river discharge; numbers indicate distance, in miles, from mouth of the James River estuary) (Hansen and Rattray 1967; reprinted by permission of the American Society of Limnology and Oceanography)

characteristics with small stratification. In Type 4, salt-wedge estuaries, the stratification is still greater, with an upper layer flowing out over a bottom layer which penetrates upstream in the form of a wedge.

Characteristics of these estuary types are most conveniently summarized in terms of the diffusive fraction of the total upstream horizontal salt flux. Lines of constant diffusive fraction of the horizontal salt flux are shown on the stratification-circulation diagram in Figure 12. When the net flow is outward at all depths, $v = 1$, and the upstream salt flux required to balance the net outward advective flux of the mean discharge is totally diffusive. At the other extreme is shown the 0.01 curve in which 99 percent of the upstream salt flux is due to the density circulation with higher salinity water flowing upstream at depth and lower salinity in the outward-flowing surface water.

With the help of Figure 12 the placement of a particular estuary on the stratification-circulation diagram immediately gives insight into the processes taking place and how they likely would influence the distribution of any substance within the estuary. The location of a particular estuary on this diagram from a knowledge of the external factors which influence it can only be done based on empirical evidence from known estuaries.

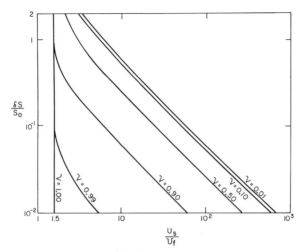

Figure 12. Fraction of horizontal salt balance by diffusion, as a function of salinity stratification and convective circulation in a rectangular channel (Hansen and Rattray 1967; reprinted by permission of the American Society of Limnology and Oceanography)

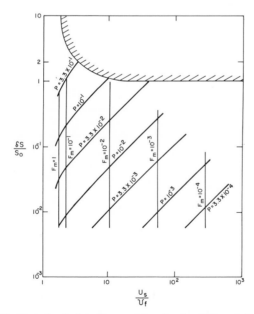

Figure 13. Stratification-circulation diagram showing isopleths of the bulk parameters F_m and P (Hansen and Rattray 1967; reprinted by permission of the American Society of Limnology and Oceanography)

Within the limits of presently available observational data, approximate relationships shown in Figure 13 can be used. Here are plotted isopleths of the bulk parameters $F_m = \dfrac{U_f}{U_d}$ and $P = \dfrac{U_f}{U_t}$ where U_f is the river discharge per unit area of cross section, U_t is the RMS tidal current speed, and U_d is the "densimetric velocity," which is equal to $\sqrt{g\,D\,\dfrac{\Delta\rho}{\rho}}$ where $\Delta\rho$ is the density difference between the river water and sea water, g is acceleration of gravity, and D is the depth of the estuary. The effects of winds have not been included herein. Although some theoretical results are available for the effect of wind on the circulation, observational data are insufficient to permit their empirical verification.

Data points from several estuaries are plotted on the stratification-circulation diagram, Figure 11. These data are specific to particular places and sets of conditions in the respective estuaries and vary from section to section and from time to time. Entire estuaries under any given set of conditions are characterized in the stratification-circulation diagram by a line rather than a point, which may cross class boundaries, putting different parts of the estuary into separate classes. Data from two stations in the James River show two points on such a line. Similarly, a section of an estuary must be expected to change its characteristics in response to external factors, such as seasonal changes and fresh-water input. The points for the Mississippi River, Columbia River, and Silver Bay illustrate the nature of these changes. In the last two cases, the changes in runoff have caused these sections of the estuaries to change classes.

A comprehensive set of data (Pritchard, in Lauff 1967) for the James River estuary illustrates the typical behavior for an estuary of Type 2. Figure 14, showing the vertical profile of mean ebb and flood currents, yields the results shown in Figure 15 for the vertical profile of the net nontidal velocity, with a typical seaward flow in the upper layer and a landward flow in the lower layer, and the depth of no net motion somewhat shallower than half depth. Figure 16 shows the mean vertical salinity profile with salinity increasing with depth. The data permitted a quantitative determination of the vertical exchange processes. Figures 17 and 18 show, respectively, as functions of depth, the mean vertical velocity and the vertical turbulent flux of salt. The turbulent flux distribution along with the mean vertical salinity profile permit a determination of the vertical eddy diffusivity as a function of depth, as shown in Figure 19. Eddy coefficients typically decrease to zero at the top and bottom. The mid-depth minimum is caused by the increased static stability from the halocline at this depth.

Conditions in a fjord can be illustrated by the low and high runoff conditions in Silver Bay. Figure 20 shows temperature, salinity, and velocity pro-

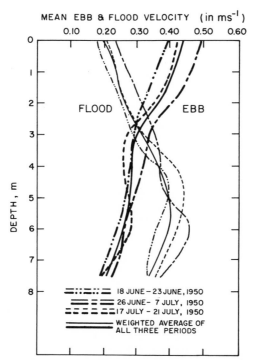

Figure 14. Vertical profile of mean ebb and flood currents in the James River estuary (D. W. Pritchard in Lauff 1967; reprinted by permission of the author and the American Association for the Advancement of Science)

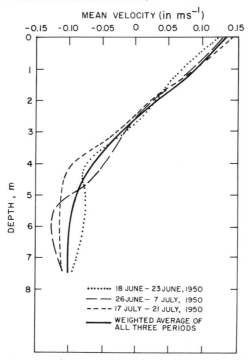

Figure 15. Vertical profile of net nontidal velocity in the James River estuary. Net flow is seaward in the upper layer (positive values) and up-estuary in the lower layer (negative values) (D. W. Pritchard in Lauff 1967; reprinted by permission of the author and the American Association for the Advancement of Science)

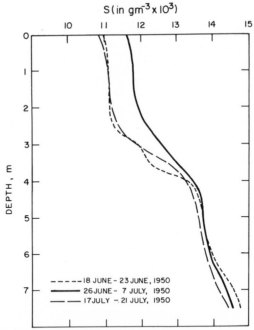

Figure 16. Mean vertical salinity profile in the James River estuary (D. W. Pritchard in Lauff 1967; reprinted by permission of the author and the American Association for the Advancement of Science)

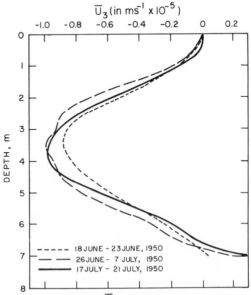

Figure 17. Average vertical velocity ($\overline{U}_3$) as a function of depth for the James River estuary (D. W. Pritchard in Lauff 1967; reprinted by permission of the author and the American Association for the Advancement of Science)

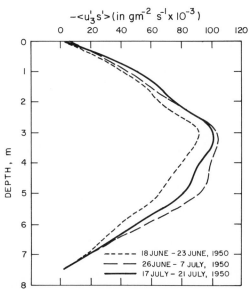

Figure 18. Vertical eddy flux of salt (U'_3S') as a function of depth for the James River estuary (D. W. Pritchard in Lauff 1967; reprinted by permission of the author and the American Association for the Advancement of Science)

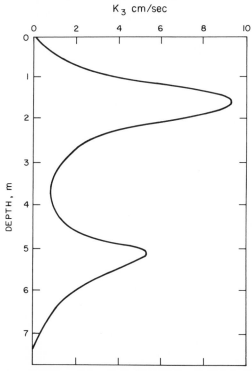

Figure 19. Vertical eddy diffusivity (K_3) as a function of depth for the James River estuary (D. W. Pritchard in Lauff 1967; reprinted by permission of the author and the American Association for the Advancement of Science)

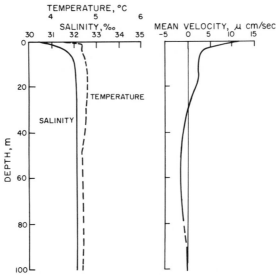

Figure 20. Temperature, salinity, and velocity profiles, Silver Bay, Alaska (March 1957) (Maurice Rattray in Lauff 1967; reprinted by permission of the author and the American Association for the Advancement of Science)

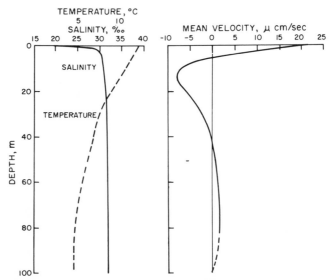

Figure 21. Temperature, salinity, and velocity profiles, Silver Bay, Alaska (July 1956) (Maurice Rattray in Lauff 1967; reprinted by permission of the author and the American Association for the Advancement of Science)

files during the low runoff period of March, while Figure 21 shows the corresponding profiles during the high runoff period in July. During the low runoff period, the circulation consists of two layers, an outflow in the surface layer and an inflow at depth. This is the case of small stratification. During the high runoff period, with large stratification, the depth of the outflow is decreased with an intermediate inflowing layer immediately underneath. At great depths there is again a small velocity net outflow. There is a shallow strong halocline with a very low surface salinity.

REFERENCES

Cameron, W. M., and D. W. Pritchard
    1963   "Estuaries." Pp. 306–24 in M. N. Hill, *The Sea,* vol. 2. New York: John Wiley & Sons.
Hansen, D. V., and M. Rattray, Jr.
    1967   "New Dimensions in Estuary Classification." *Limnology and Oceanography* 11:319–26.
Lauff, G. H., ed.
    1967   *Estuaries.* American Association for the Advancement of Science, Publication 83. Washington, D.C.
Leendertse, J. J.
    1970   *A Water-Quality Simulation Model for Well-mixed Estuaries and Coastal Seas.* Volume 1: *Principles of Computation.* The Rand Corporation, Memorandum RM-6230-RC.

Global Distribution
of Organic Production
in the Oceans

KARL BANSE

ALL life depends on organic matter made by plants from carbon dioxide, water, and nutrient salts (primarily nitrogen compounds, phosphorus, and potassium), thereby binding energy provided by sunlight. Light sufficient for plant production penetrates into the ocean for several tens of meters, up to a maximal depth of 150 meters. Because the average depth of the oceans is almost four thousand meters, life in most of the water column depends on organic matter supplied from the upper layers. Within the upper layers, annual plant production figures can be shown to be primarily related to the supply of nitrogen and phosphorus; carbon dioxide and potassium never limit production in the sea. Temperature and salinity as such have little effect on plant production figures: very high production rates occur between ice floes, and under otherwise identical conditions, lake water is as productive as sea water. Seasonal production rates in the sea are determined by the available light rather than the nutrient salts in the temperate zones during winter, and poleward from them during ever greater parts of the year.

Most of the open-ocean plants are the microscopic unicellular algae, known as phytoplankton. Outside of shallow water areas, most of the phytoplankton dies from being filtered off and eaten by the herbivorous zooplankton, animals that are a few tenths of a millimeter to a few millimeters long. Some of the ingested plant matter is not assimilated but is voided (Fig. 22). This fraction may be eaten incidentally by other animals or else, together with the material not ingested, it is broken down by bacteria or buried in the seabed. The greater part of the assimilated matter is oxidized in the animals (through respiration), primarily supplying energy for maintenance and

Karl Banse is professor of oceanography at the University of Washington. This paper is Contribution No. 684, Department of Oceanography, University of Washington.

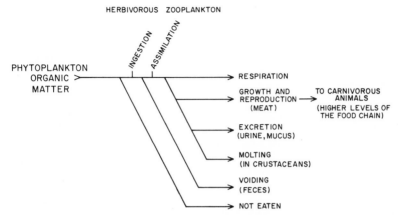

Figure 22. Scheme of the organic matter utilization by herbivorous zooplankton

movement. While supplying energy, respiration breaks down organic matter so that carbon dioxide and nutrient salts are excreted and can be re-used by the plants. Much of the organic matter not being respired is put into growth and reproduction, but occasionally the molting of crustaceans may require roughly as much organic matter over the life span of the animals as does growth (Lasker 1966; but see also Mullin and Brooks 1967 for a contrary opinion).

Previously it was reckoned that over a certain period of time the organic matter put into growth and reproduction by herbivorous plankton is 10 percent of the plant production available to them. Recent measurements in the laboratory suggest a figure of 15 to 20 percent. Generally, less animal tissue is being produced per unit of plant matter at low rates of plant production than at high rates (cf. Riley 1963; Strickland, this volume). During periods of meager food supply, the ingested organic matter has to cover the animals' needs for respiration (maintenance) first before it can be put into growth; in fact, there are seasons when no growth may occur at all. Therefore annual phytoplankton production figures are not particularly useful for estimating annual production by animals. Details of the seasonal cycle of phytoplankton production (even within the temperate zones) and of life habits of zooplankton become very important (Heinrich 1962).

Some exploited fish species feed partially on phytoplankton, making a relatively large amount of the plant production accessible to man. By and large, however, it is the herbivorous zooplankton that makes the plant production available to larger, carnivorous animals, which again use most of the assimilated matter to fill respiratory needs. Their efficiency (as defined above) is not known from experimentation, but is put by many at 10 percent. When these animals are exploited by man, there is a corresponding loss in return of the original plant production.

THE OCEAN BEYOND THE CONTINENTAL SHELVES

In a closed ecosystem (an aquarium with a lid, stocked with nutrient salts, plants, animals, and bacteria; or a future spaceship on a long mission), the organic matter produced by plants can indeed be rather completely recycled by the animals and bacteria; for example, the nutrient salts can be used over and over. This is also true for the sea when periods of several hundred years are considered and the small amounts of organic matter buried permanently in the sediments are neglected. During shorter time spans, however, the surface layers of the open ocean beyond the continental shelves (92 percent of the area of the ocean, or 65 percent of that of the globe) are not a closed system. While the nutrient salts are tied up in nonmotile organic matter (much of the phytoplankton, or feces and other dead particulate matter), they are subject to gravity and tend to fall out of the lighted layers. Phytoplankton may sink one to five meters a day; fecal pellets may sink several tens of meters during a day. Most of this organic matter is broken down fairly rapidly in deep water so that very little reaches the deep seabed, but the nutrient salts are liberated below the lighted zone so that plants cannot re-use them immediately.

Counteracting gravity and maintaining fertility of the oceans are mechanisms that return deep water to the surface; the supply of nutrient salts from rivers or of nitrogen compounds from the atmosphere is of little importance. The prime mechanisms are upwelling and upward vertical mixing. Upwelling refers to wholesale upward movement of water from depths of up to 200 or 300 meters due to horizontal removal of the surface layer by wind and the effect of the earth's rotation on the direction of currents. Upwelling is particularly important in the tropical and subtropical seas (almost one-half of the ocean) where a warm upper layer usually seals off the deep water, cool and rich in nutrient salts, from contact with the surface the year around. Wherever in the tropical and subtropical oceans nutrient-rich waters approach the surface, very high rates of plant production, and hence of animal production including fishes and whales, result because of the large amounts of sunlight available. This pattern as found in the eastern tropical Pacific Ocean is demonstrated in Figures 23–26. The shallower the surface layer (as indicated in Fig. 23 by the depth of the thermocline where the temperature drops rapidly), the closer is the deep water to the surface. Here nutrient salts easily can enter the lighted layer, causing phytoplankton development (increase in the turbidity of the water, Fig. 24). Figure 25 shows the larger zooplankton (larger than 0.3–0.5 mm), which includes herbivorous and carnivorous forms. The sperm whales (Fig. 26) are believed to feed on squid, which live from fish which in turn feed on zooplankton.

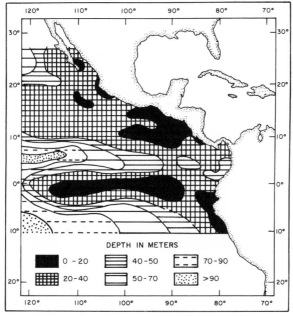

Figure 23. Depth of thermocline in the eastern tropical Pacific Ocean, October to December, 1944–56 (after Brandhorst 1958; by permission of the author and the Conseil International pour l'Exploration de la Mer)

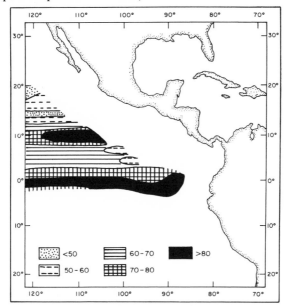

Figure 24. Particle content (in arbitrary units) of the eastern tropical Pacific Ocean, November to December 1947 (after Jerlov 1953; by permission of *Tellus*)

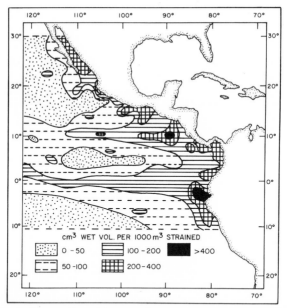

Figure 25. Zooplankton collected by nets from the upper 300 meters of the eastern tropical Pacific Ocean, October to December, 1955–56. Data in cm³ wet volume per 1,000 m³, approximately corresponding to gram wet weight per 1,000 m³ (after Brandhorst 1958; by permission of the author and the Conseil International pour l'Exploration de la Mer)

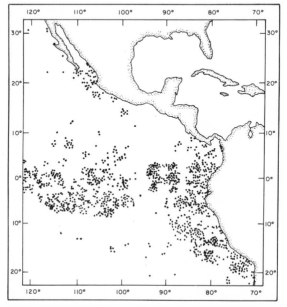

Figure 26. Position of whalers on days when one or more sperm whale catches were made in the eastern tropical Pacific Ocean in October to December, 1761–1920 (after Townsend 1935; by permission of the New York Zoological Society)

The highest daily rates of phytoplankton production observed in the sea have been found in upwelling regions. Elsewhere, in areas with warm water overlying cool water, there are always a slight upward movement and some vertical mixing related to diffusion (although not on the molecular level) which transport deep water upward. Compared with the vertical velocity of upwelling water, these processes are fairly ineffective in returning nutrient salts to the surface layer (see areas with a deep thermocline in Fig. 23).

In the temperate and subpolar zones, a warm surface layer is present during only part of the year. In autumn and winter, surface cooling leads to formation of cold dense water, which finally may sink; the same amount of deep water will rise and fertilize the upper strata. This process is called vertical mixing by convection. The depth involved varies, but it can be very large. The nutrients introduced into the upper layers may not be exploitable until spring because the light levels during winter are generally too low, but considering an entire year at intermediate latitudes, the phytoplankton production of some regions with seasonal vertical mixing approaches that of areas of lower latitudes with seasonal upwelling. The intricate interplay between vertical mixing, nutrient supply, and seasonal changes of light is theoretically fairly well understood (Riley 1963, 1965; Steele 1961). It leads to marked regional differences of phytoplankton production within the temperate zones and within any region in any given year. This holds for the seasonal distribution of production and for the annual level.

The seasonal cycle of phytoplankton in an imaginary open temperate ocean without zooplankton would consist of a dark, lean season with high nutrient-salt concentrations near the surface, followed by the so-called spring bloom of phytoplankton when enough light became available for growth. (The farther poleward the area, the later this occurs, until in the central Arctic Ocean the "spring" bloom starts in July or August and is followed by the winter.) At intermediate latitudes, after the nutrient salts would have been consumed by the spring bloom and removed from the surface layer by sinking of phytoplankton, the summer production would be low. Its level would be determined by the rate of supply of nutrients caused by rise of deep water and by vertical mixing from the diffusionlike processes. In the real ocean, with zooplankton present, recycling of nutrient salts by oxidation of plant material in the lighted layers is the other source of such salts for the phytoplankton. Generally speaking, recycling is more important the deeper the warm surface layer is; the average nutrient-salt atom may be used several times before it is lost to deep water. In the long run, however, in this situation production can be maintained only if there is a mechanism of replenishment of nutrients to the surface layer from below. In any event, phytoplankton production in the open ocean is lower in summer than in spring.

The amounts of organic matter produced by phytoplankton are usually expressed as grams of carbon per square meter (multiply by 4.047 to obtain metric tons per acre). The corresponding amount of organic matter is about twice the carbon value. Annual production in upwelling regions is several hundred grams of carbon per square meter (probably reaching five hundred grams). Poor regions in the tropical areas are well below fifty grams. In the high polar regions, the annual plant production is only a few grams of carbon. The ocean average is between fifty and one hundred grams (Ryther 1963; Steemann Nielsen 1963). An accurate world map of phytoplankton production is not yet available.

The open North Pacific off the coast of Washington and Oregon, with seventy grams of carbon per square meter annual phytoplankton production, is a poor temperate region because of hydrographic conditions that prevent deep vertical mixing during winter. In the upwelling area near the coasts of Washington and Oregon, the annual production values are likely to be 300 grams of carbon (Anderson, personal communication, 1964).

In comparison, fertile temperate sites on land (including swamps) attain very roughly fifteen hundred grams of carbon per square meter, agricultural plants reaching the level of natural communities only by careful cultivation; the production of grains by cereals is about 30 percent of their total production of organic material (Westlake 1963). Tropical terrestrial communities (including swamps) reach maximally a little more than twice the above value. Sugar cane attains maximally four thousand grams. Peak values for daily rates of natural marine phytoplankton populations are around five grams of carbon per square meter or somewhat higher. Sugar cane and maize maximally synthesize eleven grams of carbon per square meter a day (values converted from Westlake 1963). As stressed earlier, the animal production is not strictly proportional to the phytoplankton production.

Because most phytoplankton is eaten in the upper layers, and the remaining plant cells sink slowly relative to the mean depth of the ocean, there is little plant food below the upper one hundred to two hundred meters. The animals in the deeper strata, which are mainly carnivorous (besides some that are omnivorous), depend largely on the gradual transfer of organic matter from one layer to the next by feeding excursions, or on large particles sinking fast, like fecal pellets or carcasses. Because most of the organic matter eaten each time is used for respiration rather than for production of new meat, the food supply even at one thousand meters, a quarter of the average depth of the oceans, must be very low. The concentrations of animals at this depth are quite small (Fig. 27). There are no exploitable living resources below some hundred meters of depth in the open ocean.

THE SEAS ON THE CONTINENTAL SHELVES

The phytoplankton production on the continental shelves is different from that of deep water in that the seabed prevents the nutrient salts tied up in

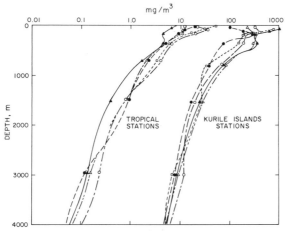

Figure 27. Vertical distribution of zooplankton collected by nets (wet weight in mg/m³ equaling g/1,000 m³) in the cool temperate northwest Pacific Ocean (right-hand family of curves) and the tropical western Pacific Ocean (left-hand family of curves) (Banse 1964, after Vinogradov; reprinted by permission of Pergamon Press Limited)

particulate matter from sinking to great depths. In fact, organic matter reaching the bottom on the shelf supports so much animal life that, in spite of the losses due to respiration, demersal fishes preying on bottom animals occur in large enough quantities to be exploitable by man. In depths shallow enough to be in the upper, warm layers, the nutrient salts continually liberated by the respiration of bottom organisms are immediately available to the phytoplankton and can be re-used several times during one season, raising the annual production figure considerably above that of the open ocean with the same concentration of nutrient salts.

The bottom environment favors bacteria much more than does the open water, and much nutrient-salt liberation is believed to be due to them. Because the activity of bacteria is particularly temperature-dependent, nutrient regeneration on the shelf must be higher in late summer than in spring. Since phytoplankton on the shelf after the spring bloom is as dependent on the rate of nutrient supply as that of the open ocean, plant production in shallow water can be higher in summer than in spring (Anderson and Banse 1963). As stated earlier, details of the seasonal cycle of phytoplankton production can be very important for animal production.

Nutrient salts supplied to the sea by river runoff are generally not important for phytoplankton production outside the estuaries, although the nutrient-salt concentration in river water can be quite high. Sea water of normal coastal salinity contains little river water per unit volume, so that the original high nutrient concentration has been diluted by sea water. However, river runoff indirectly fertilizes the coastal waters because the mixing of river water with sea water (until almost the salinity of oceanic water is reached) requires continually very substantial amounts of sea water. Thus the seawater drawn on the continental shelf by the mixing processes brings with it much of the nutrient salts consumed there by the phytoplankton.

Generalizations about the level of shelf production cannot be made, in spite of many more extended series of measurements than are available from the open ocean, and fair theoretical understanding in some regions. Outside polar and subpolar areas severely limited by light (particularly in the presence of coastal ice cover), annual figures seem to be between slightly above fifty grams and a few hundred grams of carbon per square meter, the latter values reached in unpolluted areas only where there is upwelling of deep ocean water onto the shelf (Ryther 1963; Steemann Nielsen 1963; and Figs. 23–26). The seasonal distribution of phytoplankton production is more variable than in the open ocean because of the highly variable local conditions of depth, turbidity of the water, and river runoff.

On the shallower parts of the continental shelves, plants are also found on the seabed. They are unicellular and microscopic, similar to phytoplankton, or macroscopic like kelp, red algae, and eelgrass. The production process in stands of macroscopic marine plants is similar to that on land in the sense that there is a seasonal increase of plant matter, resulting in beds. (Phytoplankton concentrations are usually quite low because the animals keep up fairly well with plant production, except for brief periods of spring blooms and in red tides.) However, contrary to the case on land, the kinds and numbers of animals feeding on plant beds in the sea are fairly small. None of the large fisheries is based on forms feeding on macroscopic algae. Some red algae are directly eaten by man, but most algae are used for industrial purposes or, along with eelgrass, as manure.

Beds of kelp and red algae are quite localized because of the absence of roots in algae, which instead require a hard substrate for affixing themselves. Large parts of the shelves are covered by sandy or muddy sediments. There is a further restriction in the vertical direction of plant distribution because of the dependence on light, so plant beds rarely occur as deep as fifty meters. The famous kelp beds of California occupy only about fifty-five square miles (according to a 1961 mapping by the United States Coast and Geodetic Survey; M. Neushul, personal communication). In cold regions, the uppermost few meters, well lighted during the summer, are razed annually by drifting ice, thus further reducing the area potentially inhabitable by plants.

Annual production rates of macroscopic benthic plants are much higher than those of phytoplankton. Several hundred to one or two thousand grams of carbon per square meter seem to be common (Ryther 1963). The contribution of benthic macroscopical plants to the total marine plant production is not well known, but is likely to be between 1 and 10 percent (the latter value from Ryther). Therefore, most of this review has concentrated on the plants of the open sea, the phytoplankton.

REFERENCES

Anderson, G. C.
1964   "The Seasonal and Geographic Distribution of Primary Productivity off the Washington and Oregon Coasts." *Limnology and Oceanography* 9:284–302.
————, and K. Banse
1963   "Hydrography and Phytoplankton Production." In M. S. Doty, ed., *Proceedings of the Conference on Primary Productivity Measurement, Marine and Freshwater* (held at the University of Hawaii, 21 August–6 September 1961). Washington, D.C.: U.S. Atomic Energy Commission, Technical Information Division, Ref. TID-7633, Biology and Medicine.
Banse, K.
1964   "On the Vertical Distribution of Zooplankton in the Sea." In M. Sears, ed., *Progress in Oceanography*, 2:53–125. New York: Pergamon Press.
Brandhorst, W.
1958   "Thermocline Topography, Zooplankton Standing Crop, and Mechanisms of Fertilization in the Eastern Tropical Pacific." *Journal du conseil permanent international pour l'exploration de la mer* 24:16–31.
Heinrich, A. K.
1962   "The Life Histories of Plankton Animals and Seasonal Cycles of Plankton Communities in the Oceans." *Journal du conseil permanent international pour l'exploration de la mer* 27:15–24.
Jerlov, N. G.
1953   "Studies of the Equatorial Currents in the Pacific." *Tellus* 5:308–14.
Lasker, R.
1966   "Feeding, Growth, Respiration and Carbon Utilization of a Euphausiid Crustacean." *Journal of the Fisheries Research Board of Canada* 23:1291–1317.
Mullin, M. M., and E. R. Brooks
1967   "Laboratory Culture, Growth Rate, and Feeding Behavior of a Planktonic Marine Copepod." *Limnology and Oceanography* 12:657–66.
Riley, G. A.
1963   "Theory of Food-Chain Relations in the Ocean." In volume 2 of M. N. Hill, ed., *The Sea*. New York: John Wiley & Sons.
1965   "A Mathematical Model of Regional Variations in Plankton." *Limnology and Oceanography* 10(Suppl.):R202–15.

Ryther, J. H.
  1963   "Geographical Variations in Productivity." In volume 2 of M. N. Hill,
         ed., *The Sea*. New York: John Wiley & Sons.
Steele, J. H.
  1961   "Primary Production." In M. Sears, ed., *Oceanography*. American As-
         sociation for the Advancement of Science, Publication 67, pp. 519–38.
Steemann Nielsen, E.
  1963   "Productivity, Definition and Measurement." In volume 2 of M. N.
         Hill, ed., *The Sea*. New York: John Wiley & Sons.
Townsend, H. C. H.
  1935   "The Distribution of Certain Whales as Shown by Logbook Records of
         American Whaleships." *Zoologica* 19:1–50.
Westlake, D. F.
  1963   "Comparison of Plant Productivity." *Biological Reviews* 38:385–425.

The Marine Food Chain and Its Efficiency

J. D. H. STRICKLAND

THE practical reasons for wanting to know the nature of the marine food chain, or better, the marine food web, and its efficiency are associated with fisheries. Even though most of the fish harvesting of the world results from hunting rather than cultivating, it is becoming vital, with many fish stocks, to know the sustainable yield, or how much can safely be taken from the sea each year. We also need to know which factors affect the abundance of each annual class, such as the critical timing of the feed available to fish larvae or the onset of heavy predation. As aquaculture increases and more fish farming is undertaken in coastal waters and lakes, we will want to be able to predict yields from optimum rations fed to stock that are present at optimum densities.

The study of the standing stock of a fishery and the way it changes with time is the study of population dynamics. Much of population dynamics can be studied by direct observation and the use of mathematical models to describe what has been observed. Eventually, however, unpredictable factors will emerge and our predictive abilities will be limited. This has been the sad history in fish population dynamics over the years; many of the troubles have been traced to problems of food availability or of unexpectedly heavy predation, although factors such as disease, physiological stress, or behavior patterns also cause populations to fluctuate.

It is clearly desirable to know what eats what in the sea—that is, the food chain—and how quickly and how efficiently the eating occurs. A quantitative description of the food web involves trophodynamics (or trophic-dynamics), which is conveniently studied under two subheadings: (1) the rate of transfer of matter between each section of the food web, and (2) the efficiency with which the matter transferred at each step is used for various

Dr. J. D. H. Strickland (1920–70) was head of the Marine Food Chain Group, Institute of Marine Resources, University of California, San Diego.

stated processes. These processes include the production of young, new flesh (which itself will be transferred in another link of the food web) and the production of the energy that is used by an organism for the purpose of staying alive.

A knowledge of the food web and trophodynamics may be sufficient for predicting population dynamics but, especially with animals having highly developed nervous systems, it does not explain everything. We also need to know something about the behavior patterns of animals. Reproduction, cannibalism, and poor parenthood are not entirely a matter of food supply. However, the *effect* of behavior is adequately reflected in the corresponding efficiency with which matter is used. For example, a population in surroundings that cause poor mating will have a low efficiency in the production of fertile eggs. Although this low efficiency might well be capable of precise measurement, the measurement will not tell us why the value is low.

FOOD CHAINS

One can draw innumerable diagrams of food webs according to what facets one wishes to illuminate. Food web diagrams, often fancily illustrated, have been appearing in popular and serious texts of marine biology for a hundred years. In the version shown in Figure 28, stress has been laid on the early part of the food web and the large stock of dead organic matter that exists in the sea. The areas of the various parts in Figure 28 are proportional to what I would guess to be the relative standing stocks of various components which might be encountered in fertile coastal seawater, and the breadth of the arrows gives some idea of the relative rates at which material passes from one part of the food web to another. Figure 29 (taken from a diagram by John Steele of Aberdeen, Scotland) was originally given to illustrate the interaction of pelagic and bottom-dwelling plants and animals. In this chapter we will be concerned mainly with the free-floating plant plankton (phytoplankton), the small animal plankton (zooplankton, mainly crustaceans) which feed upon them, and the larger animal plankton and fishes which eat these herbivores. This is the most simple sort of marine food chain and the one through which passes by far the largest amount of marine food on its way to man via the edible fish.

Before the efficiency of such a system is discussed, some of the ways in which it differs from the more familiar land-based food web, of which *man* is a part, should be pointed out. First, there are more species of plants and many more species of animals on land than in the sea because the physical isolation of one habitat from another is more readily brought about on land than in water. Second, there are generally several more stages in the food web between planktonic plants and fish of a reasonable size than between grass or leaves and animals eaten by man, because the planktonic sea plants

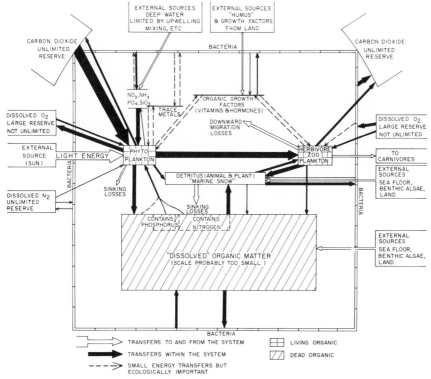

Figure 28. The food web in fertile coastal seawater

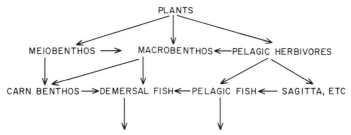

Figure 29. Feeding interactions of pelagic and bottom organisms

are very small. And third, the planktonic plants are grazed more efficiently by aquatic herbivores. By and large, terrestrial herbivores tend to be limited by their predators, whereas in the sea the amount of plant foodstuff available governs the number of foragers. This is largely an outcome of the milieu and the way it governs the mechanisms by which a planktonic predator devours its prey.

SOME DEFINITIONS AND CONCEPTS IN TROPHODYNAMICS

Critical to the whole consideration of food-chain dynamics is the concept of a trophic level or stage in the food web. The assignment, or the attempted assignment, of creatures to such a level can cause a lot of trouble, especially when dealing with omnivores. For example, an herbivorous copepod eating a diatom would normally be assigned to the second stage of the food web, the first stage being the diatom. But what happens if the same animal also eats the young of another crustacean or a speck of detritus which may have originated from a fish? Such dilemmas are not as uncommon as might be supposed in the part of the food web that we are considering. Even when studying fish, we encounter cases where a predominantly herbivorous fish switches from phytoplankton to zooplankton food.

I am convinced that the Russian biologists are correct in suggesting that the best way out of this mess is to assign each species to a separate level for each specified eating pattern. In the long run, less trouble is met if we always assign an animal to a level T. Its stated food, through any stated link in the food chain, will then always be at a level T-1, and the animal itself will fall prey to a creature which, again in this connection only, will be at a level T + 1. Numerical values are not given to T unless they are given to each precisely defined part of the food web. It is quite possible, when treated this way, for an organism to be assigned two or more different numbers for its "trophic level."

Such a treatment may sound unnecessarily formalistic, but this definition of trophic level is only one of many basic definitions, such as "food" and "efficiency," which have no common acceptance in marine trophodynamics, a state of affairs leading to chaos and confusion in the literature. Not the least of the confusions has arisen over the question of whether or not one should consider matter or energy passing along the food chain. It is convenient to consider organic molecules, initially produced by photosynthesis, as stored potential energy, the energy which can be obtained when they are oxidized to carbon dioxide and inorganic fragments such as phosphate, ammonia, and sulphate. This energy is expended in nature by a plant or animal to provide heat, to permit biochemical syntheses to take place, and to do external work. The energy content stored as food (for most systems, the heat of combustion in oxygen) changes at each step of the food chain and decreases as more and more energy is expended, eventually becoming zero when all material has been degraded to carbon dioxide (the reverse of the original photosynthesis of organic matter by plants which started the food chain). This concept of food as potential energy has been a powerful and useful concept, in that a common unit (the heat of combustion) can be used to measure the "food" at all points in a food chain.

However, this heavy emphasis on potential energy has its drawbacks. First of all, it is difficult to measure heats of combustion, and few people even try. One should be wary of using concepts in science that involve measurements that are never made. Second, most of the time for most of the small portions of the food web which generally concern us, we are not much interested in energy. We rarely try to get work out of our copepods or even our dolphins. We are more concerned about food.

Here is another term requiring definition. Dictionaries are not much help. I think it is fair to say "food" is any organic or inorganic molecule that can be assimilated or metabolized by some organism *in the food web being considered.* When thinking about food chains, it may be too restrictive to limit the definition of food at any given step just to the organic matter assimilable by the organism under consideration. The unassimilable matter in the diet of one animal may well become a major food source in the diet of another animal scavenging the excretion products of the first. (Plant "food" by this definition is nearly all inorganic, although traces of organic matter are normally required.) If the right type of metabolism meets the right type of food molecule, then some of the food transferred at a given stage of the food web may, indeed, be used as a source of energy; but this is only one of its many possible uses.

The following statements summarize what I believe is a useful way to view trophodynamics. (1) Matter is transferred from one stage of the food web to another. (2) A part of the food entering an animal will be used to produce other food (new flesh or excretory products) and during this process it may, and generally will, change its average molecular composition. This new food is available for other animals. (3) A part of the food entering an animal will be used as an energy source (generally as a result of oxidation processes) and, after this, most of it will be useless for other animals. It requires the intervention of a plant, with its photosynthetic apparatus, before the material is again of much value to an animal. (4) By virtue of the photosynthetic process, the food entering a plant is of a basically different nature from the food entering an animal and should be considered separately.

The advisability of clearly distinguishing between the energy used as such and the energy simply transferred from one organism to another as potential energy in the form of organic matter was stressed many years ago by the British ecologist MacFadyen. Of course, academic hairsplitting can take place when dealing with a few groups of specialized organisms, but the above statements are broadly applicable and useful.

Because all animal food contains a considerable amount of unassimilable inorganic matter, it is a common practice to measure food and excreta on an ash-free basis. Dry weights, carbon, or specific metabolites such as proteins can all be used according to what aspect of the food chain is being emphasized. The best single unit is probably organic carbon. If an ash-free basis is used, care must be taken to see that this rule is applied to all terms in a mass balance equation (which will be discussed later). Such a procedure neglects the transfer of vital inorganic food ingredients which must be considered for certain purposes but which are negligible in most mass and energy calculations. In some marine plants and animals, silica and calcium carbonate are important structural constituents, but the kinetics of their transfer through the marine food web is rarely studied.

Potential energy is generally expressed as heat of combustion in kilocalories per gram ash-free dry weight. Caloric value depends largely on fat content and is about 4 kilocalories/gram for an organism entirely made of carbohydrate and 9 kilocalories/gram for an oil. There does not seem to be much advantage, for most plants or animals, to be excessively oily, and values for animals and plants average about 5.6. An oily calanoid has been reported to have a value as high as 7 kilocalories/gram. Most of the marine zooplankton that have been measured have values near 5 kilocalories/gram.

Although food can be expressed in heat of combustion (energy) units, we must not, of course, conclude that this energy can all be put to useful work. Because of the convenience of expressing food in heat units and the superficial similarity between the efficiency ratio of a machine and that of a food-chain step (growth divided by food input and so forth), some terrible analogies have been made between food chains, machines, and electrical circuits. This practice has even extended to the inappropriate use of thermodynamic terminology and mathematics.

THE MASS-BALANCE EQUATION

For individual animals, conservation of matter dictates that the rate at which matter enters a level in the food web must be equal to the rate at which it leaves. This gives rise to the following simple but very basic equation for an organism or population of organisms:

$$R_T{}^F = R_T{}^G + R_T{}^D + R_T{}^E + R_T{}^O + R_T{}^M + R_T{}^R \tag{1}$$

R values are rates, all in the same units of time and matter. F indicates input of food; G represents growth of new protoplasm, and D and E, losses by defecation and excretion, respectively. For some marine animals other terms are useful: O is the loss of egg or sperm, and M is the loss of molts. (If these last two elements are still attached to the animal, the terms should be considered a subdivision of G.) G, D, E, O, and M can all, theoretically, enter another part of the food web and become food for another organism. As mentioned earlier, $R_T{}^R$ is different in that this loss of mass as carbon dioxide and associated excretory products, which gave rise to the production of energy for the animal, is no longer of major food value unless external energy is supplied. Ideally, ammonia or phosphate, associated with the food molecules that have been oxidized, should be included in the term $R_T{}^R$, along with the carbon dioxide and water, and left out of the excretion term, $R_T{}^E$. In practice, there is no way in which such a distinction can be made without using elaborate radiochemical labeling techniques.

Theoretically, a mass-balance equation can be constructed for any common measure such as weight, an element, a metabolite, or in terms of energy of combustion. In practice, data are most easily obtained in terms of dry weight or carbon. A similar equation to 1 can be written for whole populations of animals; but, in this case, if the population is "stable," $R_T{}^G$ is exactly balanced by a loss term which is equal to the rate at which animals die by predation or other causes.

Measuring potential energy is more difficult than would be imagined. Unless we know all values experimentally, including the respiratory quotient value needed for the calculation of $R_T{}^R$ and the nature of the substrate molecules involved in respiration, one is doing no more than multiplying the terms of the mass-balance equation throughout by an empirical factor. $R_T{}^F$ will be called the ingestion rate, the rate at which material passes through the digestive lining of an animal. $R_T{}^A$ will be called the assimilation rate, which is equal to $R_T{}^F - R_T{}^D$. $R_T{}^G$ has been termed the productivity of level T, but I think "production" is probably a better word.

MEASUREMENTS OF RATES IN THE FOOD CHAIN

The rate at which material is transferred from one trophic level to another should, ideally, be studied for every animal and food in the sea. For even the most restricted part of the food web this is clearly an enormous task, and work on it has scarcely commenced. Even so, it is impossible to do justice to the subject in so short an account as this one.

The most commonly measured rates are those of feeding, respiration, and growth, and these are interrelated because the assimilated food that does not get degraded by respiration must result in growth, except for small losses due to excretion. The three main variables to be considered are the nature of the food, food concentration, and water temperature, although other factors such as the population density of the feeding animals and their acclimatization to the physical environment may be important.

As far as the more common marine and fresh-water animal plankton are concerned, common-sense considerations seem to apply. The construction of an animal's mouth parts and forelimbs is a good indicator of whether the animal hunts and catches its prey or else sweeps particles into some sort of filtration apparatus. (Such filtering does not, of course, imply that some hunting cannot or does not take place as well.) The size of the esophagus and catching equipment indicates the upper and lower limits of food size. With food particles near the upper size limit, one would not expect rough or spiky particles to be liked by the animal, and this appears to be the case. (Although common sense seems to be a good guide to the feeding of planktonic crustaceans, one has to be somewhat wary of depending on it. It has recently been shown that cells of the swimming plant *Dunaliella* actively swim into the mouth of a Foraminifera—not a very logical pattern of behavior.)

The size of food particles seems to be one of the most important factors. An animal may feed on food particles of an unsuitable size but much less efficiently than if a particle size is optimum, and the animal will switch food if a more suitable size becomes available in reasonable concentrations. Rates of feeding generally increase if the concentration of food particles in the water increases, but eventually the rates reach some near-constant saturation or plateau value. These rates are usually not reached, however, until particle concentrations are higher than normally found in seas and lakes.

Animals have been considered to act like little constant-displacement filter pumps, and their feeding has been expressed as a "filtration rate," generally as milliliters of water filtered per animal per twenty-four hour day. For a small, planktonic, herbivorous crustacean like *Calanus,* the typical rate might be 50 to 200 milliliters per twenty-four hours. Filtration rates are obtained, experimentally, by noting the decrease of cell numbers in a given volume of water at a given time and making a certain type of calculation based on an assumed exponential drop of food concentration with time. (This relationship seems to have been rarely tested before being used in an experiment.)

A filtration rate described in this way is certainly one of the easiest things to measure. Early workers made the assumption that this rate would be a constant characteristic of an animal, independent of the nature of its food or the concentration of this food in the water. However, it is becoming obvious that this is not strictly the case, and the time has come when we should express feeding in some other terms. It is really no more difficult to express feeding rates as a function of food concentration than to quote a filtration rate. If the feeding rate is directly proportional to the concentration of food in the water, both treatments become identical; but one makes no implication as to the *mechanism* of feeding. If either the feeding rate or filtration rate depends in a nonlinear manner on the standing stock of feed, this must be determined experimentally.

It is of basic interest in fieldwork to see if the estimated R_T^F values are as great as estimated R_T^R values; if not, we know that either the animal cannot survive or our estimate of one or the other rate is erroneous. For example, a Danish zoologist has remarked on the constancy of the following ratio for the marine zooplankton, the value of which is generally fifteen:

$$\frac{\text{Liters of water filtered in unit time}}{\text{Milliliters of oxygen consumed in unit time}}$$

Assuming a proportionality between R_T^F and filtration rate, it can easily be calculated that the water should have at least 30 milligrams of carbon per cubic meter of food present in it in order to maintain the zooplankton. This corresponds to $0.5-1$ $\mu g/l$ of plant chlorophyll, which is about the level one would expect to find in the waters from which the data were derived. In less rich waters, either respiration rates must be lower, filtration rates higher, or food must be captured by the animals in some aggregated form which we do not yet understand.

We still need a commonly accepted method for expressing rates of feeding, growth, and respiration. For a rate, it would be logical to use a measure with dimensions of reciprocal time, a suitable unit for most of the rates of plankton development being $(\text{day})^{-1}$. This leads to the following expressions where W_T is the mass of an animal, and F_T, R_T, and G_T are feeding rate, respiration rate, and growth rate constants, respectively, in units $(\text{day})^{-1}$:

$$R_T^F = F_T \times W_T \text{ or } F_T = \frac{R_T^F}{W_T} \tag{2}$$

$$R_T^R = R_T \times W_T \text{ or } R_T = \frac{R_T^R}{W_T} \tag{3}$$

$$R_T^G = G_T \times W_T \text{ or } G_T = \frac{R_T^G}{W_T} \tag{4}$$

Rates must be measured with time units of a day, with care being taken to see that the same measure of mass is used in the rate expressions as is used to characterize the animal. Such constants enable one to compare whales with shrimps. But they will not, of course, have the same value at all times for the same animal. In particular, there is a tendency for all three constants to be greatest for young and vigorously growing animals. When an animal reaches its breeding age, much of the growth is generally channeled into the production of reproductive tissue. Individual feeding or respiration rates are of little interest, and only when growth, respiration, feeding, breeding, and other rates are measured simultaneously do we get a good picture of the trophodynamics of an animal or population of animals. Good data of this nature are sadly lacking.

By recalculating results reported in the literature, one obtains values for G_T which range from as great as 0.5 per day for young crustaceans at high temperatures to about 0.05 per day for adults at low temperatures. Feeding rate constants can exceed 1.0 for young exposed to suitable food, but are more generally in the range of 0.1 to 0.3. Respiration rate constants are generally less than 0.1; but if an animal is in warm water to which it is not acclimated or if it is unduly active, values may be greater than 0.5 for short periods.

TROPHODYNAMIC EFFICIENCIES

We have so far said nothing about the efficiency of the food chain. One can learn nothing from efficiencies that one cannot learn from knowing the various rates with which matter is moved about the food chain; but, if an efficiency is more constant than the rates which are used to calculate it, it can have predictive value.

Efficiencies of Processes Taking Place with Single Animals

Any of the rate terms in equation 1 can be expressed as a fraction of any other, but we are normally concerned with the efficiency with which food is used. To calculate assimilation efficiency by using the symbols in equation 1, we have:

$$E_{T^A} = \frac{R_T{}^F - R_T{}^D}{R_T{}^F} = \frac{R_T{}^A}{R_T{}^F} \tag{5}$$

One of the few useful generalizations that we seem able to make about the trophodynamics of the plankton is that the assimilation efficiency is high. An animal feeding on reasonably suitable food will assimilate from 50 to 80 percent of the organic content of this food. There is some evidence that the efficiency of assimilation of the organic fraction decreases if there is much inorganic matter. This is a way of saying that an animal has some difficulty in absorbing organic matter when it is buried in inorganic material, such as the juice inside a silica-covered diatom.

The Russian planktologist Constantin Beklemishev has suggested, mainly on the basis of his field observations, that the efficiency of assimilation decreases when the zooplankton feeds in waters with high concentrations of particulate food (greater than about 400 milligrams of carbon per cubic meter). Feces are then supposed to become progressively richer in undigested matter and progressively more useful to other members of the food chain. This suggestion of "superfluous feeding" has received little support from experimental data obtained under laboratory conditions and it may have little general applicability—although, of course, heavy feeding will produce a heavy crop of feces. This type of feeding will still have trophodynamic significance even if the food content per unit weight of feces does not increase.

The following equations, where $E_T{}^{G(G)}$ is the gross growth efficiency and $E_T{}^{G(N)}$ is the net growth efficiency (known as first order and second order growth efficiencies by some workers), are used to calculate growth efficiency. The superscript $(T-1)$ indicates feeding or assimilation of material from trophic level T — 1.

$$E_T{}^{G(G)} = \frac{R_T{}^G}{R_T{}^{F(T-1)}} \tag{6}$$

$$E_T{}^{G(N)} = \frac{R_T{}^G}{R_T{}^{F(T-1)} - R_T{}^D} = \frac{R_T{}^G}{R_T{}^{A(T-1)}} \tag{7}$$

It is useful to distinguish between the growth efficiency of an organism over a short period (growth efficiency at age i [$E_T{}^{G(G)}$] and over an organism's entire life span up to the age of i. Growth *to* age i ($E_{Ti}{}^{G(G)}$) can be expressed as

$$E_{Ti}{}^{G(G)} = \frac{\text{energy content or amount of organic matter in animal at age } i}{\text{content of egg + content in food consumed from birth to age } i} \tag{8}$$

$E_T{}^{G(G)}$ will probably decrease with the increasing age of an animal, although recent work by Dr. M. M. Mullin in my laboratory at San Diego has shown this is not the case during the *developmental* stages of two marine copepods. $E_{Ti}{}^{G(G)}$ values in the literature rarely exceed 0.05 to 0.1 for adults, but can be 0.3 or more for young. $E_T{}^{G(N)}$ values will, of course, be up to twice the $E_T{}^{G(G)}$ values, and figures exceeding 0.9 have been reported. In recent work with copepods Mullin found $E_T{}^{G(G)}$ little affected by temperature or the nature of the food, a valuable generalization if it is borne out by further work on other similar animals.

It is interesting to consider plant growth as an intake of food in the form of radiant energy being turned into plant cellular substance as potential energy. The maximum "growth efficiency" $E^{G(N)}$ is then about 35 percent, although $E^{G(G)}$ will be considerably less since not all the light falling on a plant cell (fed to it) is absorbed (assimilated), some being scattered or lost by fluorescence. If we consider not the photons falling on the plant cell but the total number of photons falling onto the sea surface, the "growth efficiency" has a maximum theoretical value of only about 4.5 percent, and the maximum likely to be found in nature would scarcely ever exceed 1 percent, being more typically 0.01 to 0.1 percent. (Plant growth in terms of matter, rather than energy, has, by definition, an $E^{G(N)}$ value of unity, or is very close to it.)

Efficiencies of Processes Taking Place
with Populations of Animals

Whole populations of animals of the same species, as well as individual animals, can have growth and other efficiencies, obtained from the ratios of terms in equation 1. The values for populations are more ecologically significant but more difficult to measure and interpret, as the population will be made up of animals of all ages, each one of which may have different efficiencies according to its stage of development. In natural communities, however, we rarely find all animals of the same species. Organisms are both eating and being eaten, so that we must examine the question of how efficiently matter gets transferred along the food chain from one population of organisms to another.

Transfer efficiency of an organism at trophic level T. This term is also known as "ecological efficiency," but I consider the word "transfer" more descriptive. Put mathematically in the simple if cumbersome notation we have been using, this efficiency can be written:

$$E_T^T = \frac{R_{T+1}^{F(T)}}{R_T^{F(T-1)}} = \frac{I_{T+1}}{I_T} = \frac{\text{Rate of predation of T}}{\text{Rate of predation by T}} \qquad (9)$$

I_T is the intake rate at level T and can be used to replace $R_T^{F(T-1)}$ (the rate of feeding of a whole population) for convenience in writing. This efficiency shows the yield to a predator (at level $T + 1$) of food, compared with the corresponding yield of food (at level $T - 1$) to its prey. The transfer efficiency is a measure of the ability of an animal at level T (a copepod, for example) to transfer the material in plants or animals at level $T - 1$ (such as phytoplankton species) to an animal at level $T + 1$ (for example, a herring).

E_T^T differs basically from $E_T^{G(G)}$ in depending on the number of animals at level T which are actually eaten by the organism at level $T + 1$. If the animals at level T were not eaten, E_T^T would be zero no matter how much these animals were to grow. E_T^T says nothing, *per se,* about the growth efficiency of populations at level T for $T + 1$ or the standing stocks at any trophic level.

If the population is stable in number and age distribution, one can write I_T in terms of the standing stock of organisms at level T (S_T) as:

$$I_T = m_T \times S_T \qquad (10)$$

and therefore:

$$R_{T+1}{}^{F(T)} = I_{T+1} = m_T \times S_T \times E_T{}^T \qquad (11)$$

The proportionality constant, m_T, which should formally be written $m_T{}^{(T-1)}$, has been called the maintenance cost of the population T. It is the population equivalent of the constant F_T (see equation 2). From equations 6, 9, and 10 it follows that *in a stable population:*

$$\frac{S_{T+1}}{S_T} \text{ (ratio of standing stocks)} = \frac{R_{T+1}{}^G}{R_T{}^G} \text{ (ratio of productivities)}$$
$$\times \left[\frac{m_T}{m_{T+1}} \times \frac{E_T{}^{G(G)}}{E_{T+1}{}^{G(G)}} \right] \qquad (12)$$

There are very few values for $E_T{}^T$ available for the plankton, despite the obvious importance of this efficiency, although Dr. L. Slobodkin has suggested that this efficiency will be found to approach a mean of 0.1 when sufficient data are at hand. Much of the present data for aquatic environments have been obtained in rather artificial situations where an experimenter rather than a natural predator has removed the organisms at level T.

It will be seen from equation 12 that standing stocks at either end of a link in the food web are not necessarily proportional to the ratio of productivities, as is often supposed, unless the m and E^G values for the animals at either end do not change. The bracketed term in equation 12 is not easy to evaluate, but considering equations 4 and 10, it will be seen to be equal to the ratio $\dfrac{G_T}{G_{T+1}}$ where these are the *population* growth rates. This leads to the identity:

$$\text{Ratio of standing stocks } (\frac{T+1}{T}) = \text{Ratio of productivities } (\frac{T+1}{T}) \times \frac{G_T}{G_{T+1}}$$
$$(13)$$

The last term will have a value somewhere between about 10 and 20 when considering marine phytoplankton and the animals which graze them. The ratio of the standing stocks of phytoplankton and the feeding upon them by zooplankters is often found to be near to unity when we consider the whole water column in the ocean. Thus the productivity of the zooplankton may often be about 5 to 10 percent of the productivity of the plants in the next lowest trophic level. The ratio of standing stocks, however, is found to depend on the mean concentration of plant cells in the water: there are more zooplankters relative to phytoplankters, when the concentration of the phytoplankton is greater. The ratio of productivities at two adjacent levels probably causes the number of plant cells in the water to increase as both the ratio of standing stocks increases and the G_{T+1} increases with little corresponding rise in G_T values for the plants.

My colleague, Dr. M. B. Schaefer, by assuming a world productivity of phytoplankton of 2.10^{10} net tons of plant carbon, was able to set limits to fish productivity, assuming this was all at a third trophic level. He assumed various ratios of productivities between adjacent levels. Assuming a mean of 0.1 for these (which Schaefer thinks conservative), the world fish productivity would be at least four to five times the present catch.

Acceptance index of an organism at trophic level T. An organism at level T that feeds on a species which, by definition, will be labeled T — 1 may, and probably will, also eat other food. This is usefully measured by a "preference" or acceptance index, A_T^{T-1}, which can have values from zero to one and which is defined by:

$$A_T^{T-1} = \frac{R_T^{F(T-1)}}{R_T^{F(total\ food\ eaten)}} \tag{14}$$

Recently it has been possible to calculate A_T values, again from the work of Mr. Mullin and his associates. *Rhincalanus nasutus,* when presented with a mixture of diatoms of various sizes as well as small animals (*Artemia* larvae), showed distinct preferences which varied with the developmental stage of the copepod. This typical "herbivore" when fully developed had an A_T as high as 0.4 for animal food. To make a complete inventory of transfer efficiencies for an animal, the A_T values for all food likely to be encountered should be known.

Food-chain efficiency of an organism at trophic level T. This efficiency is discussed a lot by trophodynamicists but is used very little. It is defined as:

$$E_T^F = \frac{I_{T+1}}{R_{T-1}^G} = \frac{\text{Rate of predation of T}}{\text{Rate of food }supply\text{ to T}} \tag{15}$$

(Note that the denominator here is the rate of food supply or growth, *not* the amount actually consumed by the animal at level T.) The food chain efficiency *per se* says nothing about whether T — 1 is a significant part or not of the diet of *T,* although this is implied, and I am not sure if this efficiency is particularly useful either in theory or in practice, although it gives some measure of the ability of a population to cope with the task of keeping down the population at the next lowest level. $E_T{}^F$ values will clearly be less than $E_T{}^T$ values.

If all phytoplankton growth were to be cropped by zooplankton (a common occurrence), an efficiency not unlike the food chain efficiency would be obtained from the following ratio, which would have a value generally in the range 0.01 to 0.1 percent:

$$\frac{\text{Amount of plant produced (and eaten) in unit time (in energy units)}}{\text{Amount of radiant energy arriving at the sea surface in unit time}}$$

CONCLUDING REMARKS

We have, so far, said nothing about the most practically useful efficiency of the lot, that with which a given organism can be cropped from a population by artificially increased predation, the sort of predation which occurs when fish are harvested from the sea by man. The mathematics to describe such a state of affairs has been written, but the practical application to plankton populations is almost nonexistent and there seems little point in enlarging on the subject in this introductory account. The material I have set forth above is, I must confess, not of any great originality but has, I hope, outlined the problems we face when studying the early part of the food web and the efficiency of even its most straightforward pathways.

At this juncture the most sadly lacking aspect is the accumulation of good data. The field abounds with armchair theoreticians willing and able to squeeze, often with great ingenuity, the last ounce of approximation from already inadequate data; but there are few workers willing and able to go to the laboratory and out to sea to make good measurements within a simple and well-defined framework of trophodynamics. Great problems of methodology remain, especially those concerned with the adequate sampling of such a nonstatic environment as the sea.

The ultimate goal, to know every definable rate and efficiency for every step in the food web, is, obviously, unattainable in the foreseeable future, and one may question whether it would ever be worth the effort to accumulate such data. I foresee the main impetus to the study of marine trophodynamics coming when, eventually, we turn to marine aquaculture in a serious way and economists will breathe down our necks insisting that we find the efficiencies of all feasible pathways to a desirable end product so that the most profitable one can be used. As a young man working in the chemical industry, I was impressed by how effectively the corresponding approach stimulated chemists to find out more about the efficiency of the processes they used. Perhaps a new generation of biologists will one day pave the way for the first edition of the Trophodynamicists Data-Handbook.

Food from the Sea and Public Policy

WILBERT McLEOD CHAPMAN

IF there is anything that can be said definitely about the fish business, it is that every citizen above the age of about six is expert in it. No molecular biologist, mathematician, or nuclear physicist is deemed too abstruse or theoretical in his training to be barred from a presidential scientific advisory committee panel advising the president and the government at large on how to harvest the ocean. No fisherman on the dock hesitates to declare how ignorant scientists are of fish and the ocean. In between, there are few dentists, doctors, truck drivers, lawyers, or common laborers who will not admit to a modest knowledge of fish and the ocean and how they should be managed. Accordingly it is with some trepidation that I offer my own thoughts on the management of ocean resources.

THE CURRENT HARVEST OF LIVING AQUATIC RESOURCES

The world catch of living aquatic resources, excluding whales, for selected years over the past century or so has been about as follows (in millions of metric tons):

1850	1.5 to 2.0
1900	4.0
1913	9.5
1930	10.0
1938	20.5
1950	20.2
1960	38.0
1965	53.3
1966	56.8

These figures are far from precise, but they are among the least controversial aspects of the fish business. The estimates of Moiseev for the period

Dr. W. M. Chapman (1910–70) was associated with the Van Camp Seafood Division of the Ralston-Purina Company.

prior to the end of World War II are as accurate as anyone can make. Since 1948, the Food and Agriculture Organization has been compiling the fish catches of all countries as reported to it in annual volumes; these figures are acknowledged to be understated because in a good many countries the fishery statistical system is not well developed, but it is assumed that this understatement at least has been relatively constant.

What the figures do demonstrate is the rapid and steady growth of the fisheries since the end of World War II. In 1948 the catch was 19.6 million tons, in 1966 it was 56.8 million tons. The greatest part of this expansion, and the greatest part of the harvest, consists of marine fish. Percentagewise, the composition of the world catch in 1948 was 74.9 percent marine fish, 13.2 percent fresh-water and diadromous fish, 10.1 percent crustaceans and other invertebrates, and 1.8 percent other things such as seals and aquatic plants. By 1966 these percentages were, respectively, 77.4, 13.9, 7.4, and 1.3.

In absolute terms the catch of marine fish increased from 14.69 million tons in 1948 to 43.93 million tons in 1966; of fresh-water and diadromous fish, from 2.58 million tons in 1948 to 7.91 million tons in 1966; of crustaceans, molluscs, and other invertebrates, from 1.97 million tons to 4.21 million tons; and of other things, from 0.36 million tons to 0.75 million tons. Put another way, about 91 percent of the total harvest of aquatic living resources is fish, and of this about 85 percent is marine fish. If we were to add on the number of diadromous fish that spend most of their time in the sea, as extracted from the fresh-water and diadromous sector of these statistics, then marine fish would make up a little more than 90 percent of the world fish catch.

COMPOSITION OF THE WORLD FISH CATCH

FAO divides its world fish statistics into nine broad groups of species characterized by the following examples: (1) herrings, sardines, and anchovies; (2) cods, hakes, and haddocks; (3) red fishes and basses; (4) jacks and mullets; (5) mackerels and billfishes; (6) tunas and bonitos; (7) halibuts, flounders, and soles; (8) sharks, rays, and chimaeras; and (9) unsorted and unidentified fishes. The most striking aspect of the catch statistics broken down into these groupings is the explosive growth in production of the first group (the clupeoids) in the past twenty years. The world catch of herrings, sardines, and anchovies has increased from about 4.8 million tons in 1948 to about 18.7 million tons in 1966. The most spectacular individual species gain in this group has been the anchovy, whose catch in 1948 was less than 1 million tons and in 1966 was a little more than 10 million tons. Most of these catches consisted of just one species of anchovy off Peru and northern Chile which in 1948 produced only a few thousand tons and in 1966 produced 9,621 thousand tons, or a little more than half of that year's total clupeoid catch. It represents nearly a third of the total increase in marine fish catch during the past eighteen years.

Production in the second largest group (the gadoids or codlike fishes) has increased from about 3.7 million tons in 1948 to about 7.3 million tons in 1966. It is important to note that when one removes the totals of clupeoid and gadoid fishes plus the unsorted and identified fish from these statistics, all the remaining marine fish landings amount to a little less than 10 million tons, having increased to that level from about 3.3 million tons eighteen years previous. Among those fishes that are highly noted as excellent food fishes having substantial commercial value, the landings are not very large nor have they increased much. The catch of all flat fish (flounder, halibut, sole) amounted to only 1.09 million tons in 1966 and was 790,000 tons in 1958. The world-wide catch of all tunas, bonitos, and skipjacks was only 990,000 tons in 1958 and 1.32 million tons in 1966. The catch of all crustaceans (crab, shrimp, lobster) was 1.26 million tons in 1966, having increased from 850,000 tons in 1958. The catch of all molluscs (abalones, clams, oysters, mussels, squids, octopus) was 2.07 million tons in 1958 and 2.9 million tons in 1966. The catch of all salmon, smelt, and trout was 1.18 million tons in 1966 and had been 740,000 tons in 1958. Most of the growth, and nearly half of the total, in this category was capelin (521,000 tons in 1966), used mostly for fish meal.

The lesson derived from this oversimplified look at actual fish landings is that the great increase in world fish landings has not been in the expensive delicacies from the sea, but in the cheaper forms of protein.

THE USES OF AQUATIC LIVING RESOURCES

An examination of the uses to which man puts his harvest of living aquatic resources reveals certain trends. The largest category of use is still in the fresh state, but this has been shrinking relatively over the years. In 1948, 49.5 percent of landings were used fresh, but in 1965 the figure was only 33.4 percent. Smoking, curing, and drying, which had constituted the second largest use of fish in 1948 (25.5 percent), had been reduced to third place in 1965, using only 15.5 percent of landings. In 1965, 29.2 percent of the catch was used for reduction into meal and oil, as compared to only 7.7 percent in 1948. Much note has been made of the increase in frozen fish use in the world since the war, but in 1965 only 10.9 percent of landings was used for frozen products (up from 5.1 percent in 1948). Fish used for canning grew even less proportionately, from 7.1 percent in 1948 to 9.1 percent in 1965. From this examination one may conclude that the trend in use of fish is toward reduction to meal and oil, and not direct human consumption. In 1948, 87.2 percent of landings was used fresh or processed for direct human consumption; in 1965, only 68.9 percent was used for this purpose.

FOOD POTENTIALS OF THE WORLD OCEAN

Other chapters in this volume describe the processes by which organisms suitable for human food are generated in the ocean. Accordingly I will not repeat this material, but will try to ascribe some crude numerical estimates to what is produced per year.

A recent estimate of primary production suggests that the standing stock of plants in the ocean is about 1.7 billion metric tons at any one time, and that they yield a total production of about 550 billion tons per year. The standing stock of multicellular plants (kelps, seaweeds, algae) in the ocean at any one time, and the annual production, is estimated to be about 200 million metric tons. The rest is unicellular plant plankton, and it is in this category that the astronomical quantities of living matter are produced by the ocean—by these estimates, a standing stock of 1.5 billion metric tons and an annual production of about 500 billion metric tons.

Much use is already made of the seaweeds and kelps of the ocean margin for direct human consumption and other purposes. However, I know of no reasonable estimates of this use. The FAO figures are obviously far understated: while listing a total world use of 710,000 metric tons in 1966, only 2,300 tons are attributed to the United States, whereas in California alone about 160,000 tons of giant kelp are harvested annually. No matter how much the FAO figures underestimate annual world harvest of this sort of plant, it is obvious that there is much scope for increased harvest between their estimate of less than 1 million tons used now and Bogarov's estimate of ocean annual production of 200 million tons. Furthermore, some types of seaweed are grown presently for human consumption. In Korea, laver farming produces 20,000 tons per year, and in Japan perhaps ten times that much is harvested. In southern California, chemical control of sea urchins, which eat the young thallophytes, is providing a means for beginning mariculture in this field.

For the unicellular phytoplankton that the ocean produces in such volume, I know of no way of harvesting either now or in the foreseeable future whereby the value of the product could be as great as the cost of producing it. At least at present, these tiny plants have to be concentrated by animals before they can be economically harvested and used by man.

A bold estimate of the volume of animals in the ocean at any one time and the annual production of animals by the sea has been made. The first figure is set at 32.5 billion tons, and annual production at about 56.2 billion tons. I am sure that no one takes these estimates to be anything more than crude approximations of primary and secondary production of life by the ocean, based on the best scientific evidence available, which is none too good. Nevertheless, if we took these estimates to be entirely accurate, we would still be a long way from agreement as to the volume of food the ocean can produce that is useful to man. As is the case with the phytoplankton, most of the animals in the estimates are so tiny that at present there is no method of harvesting them at a profit, or even on a break-even basis. Once again, these microscopic creatures for the most part need to be concentrated by larger animals before they can be economically harvested and used by man.

However, at this level in the food chain, animals do become importantly beneficial to man. Anchovies for the most part feed on phytoplankton. They are, from the standpoint of volume, the most important source of animal protein from the sea now used by man. In 1966, the landing was recorded as 10.75 million metric tons, and the total would more likely be closer to 11 million tons if the anchovy included in the category "various marine clupeoids" could be identified. This figure is about a quarter of all marine fish caught in 1966. In 1948 the total had been only about 200,000 tons and in 1958 was only 1,431,000 tons. Furthermore, there are quite large stocks of anchovy at several points in the world ocean which are fished either lightly or not at all, in the tropics as well as in both north and south temperate waters.

The reason for such a large anchovy population in the ocean is that these fish are, for the most part, the first ecological stage above the phytoplankton and they are much more abundant than bonito, which feed on them, for the same good ecological reason that there are more grass-eating antelopes on the Serengeti Plains than there are antelope-eating lions.

The next most abundantly taken kind of fish is the family Clupeidae, which includes herring, sardinella, sprat, menhaden, and pilchard. The herring itself is for the most part carnivorous, living on copepods and other tiny animals as does the pilchard. The herring do eat some phytoplankton, however, and the tropical sardines of the genus *Sardinella* are primarily phytoplankton feeders. Although herring and pilchard are pretty heavily fished in most of their range, large underutilized stocks of tropical sardinella, thread herring, menhaden, and other related fishes are known on both sides of Central America, West Africa, both sides of the Arabian Sea, and northwest Australia. Also at this ecological level are quite large resources of other sorts of fish and invertebrates that are used lightly or not at all, such as the smaller species of shrimp, krill (euphausids), red crab (*Pleurencodes*), deep-sea smelts, and myctophids.

It is at this second trophic layer stage that estimation of the ocean's capabilities to produce fish or food begins to break down. Graham and Edwards (1962) estimated that the ocean was able to produce considerably less than 60 million metric tons of marine fishes per year without the development of totally new catching technology and pisciculture techniques. Schaefer (1965) carefully reviewed previous estimates of future fish production levels (all of which have been low) and stated (1968), "Considering this kind of information, I have concluded that, at a conservative estimate, the world fishery production may be increased to 200 million metric tons per year, with no radical developments, such as fish farming or far out kinds of fishing gear."

But even these two estimates did not refer to exactly the same thing. Graham and Edwards were talking about marine fishes strictly; Schaefer was including all organisms supporting traditional fisheries, including crustaceans and molluscs. Both, however, were talking about "traditional" fisheries, that is, those now being prosecuted. However, I do not feel that this is a sufficiently comprehensive view of ocean food production capability. The nature of demand for and use of fish is changing quite rapidly, as can be seen, in part, by the great surge forward in production of fish meal noted above.

In 1948, world production of fish meal was about 590,000 metric tons and in 1958 it increased to 1,374,000 tons. In 1967 it was estimated that world production would be about 4.5 million tons. But this is not the only growing new use for fish. The production of Alaska pollack went from 578,000 tons in 1962 to 1,221,000 tons in 1966; while the bulk of this catch was used for fish meal, during the last few years a large part has been turned into ground-up frozen fish which is taken back to Japan and used in fish cake for direct human consumption. The catch of capelin surged up from 38,000 tons in 1964 to 521,000 tons in 1966. Again, the greater part of this has gone into fish meal, but the Japanese are now buying capelin in Norway to put into fish cake. We also have the prospect for fish protein concentrate coming into use in the near future, by which cheap protein from the sea will be incorporated into the human diet on a broad basis.

There are four principal things to be kept in mind concerning the future of fishery development. First, the nutritional value of animal protein from all fish flesh is approximately identical, whether it is anchovy in Peru that brought the fisherman fourteen dollars per ton, or bluefin tuna in Tokyo that brought the fisherman three thousand dollars per ton.

Second, the costly part of fish production is the preservation of taste, texture, and fragrance of the freshly caught fish. These qualities are delicate and evanescent in nearly all fish. Their preservation, or stabilization to a desired state, by canning, freezing, curing, drying, or irradiation, is expensive and has nothing whatever to do with preserving the nutritive value of the fish.

Third, the highly prized and costly fish on the world market, such as salmon, tuna, billfishes, cods, flat fishes, ocean perches and related percomorphs, are carnivorous animals several ecological steps up the food chain from the original plant plankton. There simply are not very many such third- or fourth-stage carnivores in the ocean, and if we are to depend upon them for producing food from the ocean, then Graham and Edwards' estimates are not too conservative.

The situation can be illustrated, if somewhat oversimplified, by this example. If one thousand tons of plant plankton off Peru were eaten by anchovy, one would not expect a yield from this of much more than one hundred tons of anchovy. The rest of the energy included in the plankton would be used in the ecological inefficiency of changing trophic levels. If the one hundred resultant tons of anchovy were eaten by bonito, one would not expect a yield of more than about ten tons of bonito. If the ten tons of bonito were eaten by yellowfin tuna, one would expect a yield of about one ton of tuna from the ten tons of bonito, or from the original one thousand tons of phytoplankton.

Fourth, the animals closest to the phytoplankton in the ecological system tend to be smaller than those in farther removed trophic levels. Most phytoplankton-eaters are microscopic in size, and relatively few are more than five inches long. When reviewing changes in trends of fish use as noted above, some account needs to be taken of the size of animals used for different purposes. The bulk of fish that are used for prime table fish, whether fresh, frozen, or canned, are ten or more inches long. The great bulk of fish used for fish-meal production, however, are from five to ten inches long, and it is in this size range that the great expansion in landings in the past ten years has occurred.

There are more animals in the sea, however, in the adult size range from one to five inches in length than there are totally in the other two size classes referred to, from five inches in length on up; and if the great expansion in ocean food production that has gone on during the last decade is to continue, it is going to be necessary to look to smaller fish, from five inches to one inch in length.

As an example, the krill of the Antarctic should alone be able to sustain an annual yield of 100 million tons (Kasahara 1966); and they belong to the size range from one to two inches in length. The highly abundant lantern fishes and deep-sea smelts, also in the one- to five-inch length range, are not now used at all by man; and the sand lances, which are now yielding about 250,000 tons per year (mostly from Denmark and Japan), are not yet fished at all in the northwest Atlantic where they appear, from egg and larvae surveys, to be the most abundant fish.

If it is taste, texture, and fragrance we desire in food from the ocean, then it is the third- and fourth-stage carnivores that we must cultivate, not expecting to enlarge greatly the present production of food from the sea. If, however, we want the nutritional value of animal protein well balanced in amino-acid content for the human body's use, then we must look to the smaller sizes of sea animals. If we take the first course, then a sustainable annual catch of 60 million tons per year may not be much too low a level to expect. If we take the second course, then I believe that a sustainable annual yield from the world ocean in the range of 2 billion tons is not too much to expect. My estimate is not incompatible with that of Schaefer.

SOME UNITED STATES VIEWS ON FOOD FROM THE SEA

There are several misconceptions about food from the sea held by the federal government, and by the general public, in the United States that have affected public policy on food from the sea.

1. Stationary market for food from the sea. It is generally held that the market for fish in the United States is poor, that consumption stays the same generation after generation, and that the salvation of the domestic fish business lies in marketing aids for the increased use of fish. Statistics show that the United States consumed 2.8 million short tons round weight of fish in 1948 and 6.2 million short tons in 1966. Instead of the per capita fish use remaining at 10 to 11 pounds per year, it was about 38 pounds in 1948 and increased to about 60 pounds in 1966. The United States stands second only to Japan (and nearly on a par with it) as a lucrative, rapidly expanding market for living resources of the sea.

2. Wastefulness of feeding fish to chickens. There is a general feeling in the United States that it is slightly immoral to feed fish to chickens when there is so much protein malnutrition in the world. However, chickens transfer animal protein into higher value animal protein (fresh meat and eggs) somewhat more efficiently than do tuna, salmon, and halibut. Peruvian anchovy cannot be harvested and used economically in the volumes now used except when produced as fish meal for animal husbandry. If not used for that purpose, they will die and revert into the web of life in the ocean unused by man, as does most of the animal protein produced by the ocean. Fish meal fed to chickens, turkeys, pigs, and calves is contributing as directly to the human diet as are tuna, salmon, or halibut. The practice of using fish meal in chicken feed is what has made chicken, turkey, and eggs cheap and widely used, because the nutritional efficiency of the bird is so much improved by this balanced diet.

3. Lack of growth in the United States fish industry. Much is made of the decadence and general backwardness of the United States fish industry. Proof is adduced by the fact that landings by United States flag vessels have stayed between 2 and 3 million tons per year for a generation and that landings by the efficient Russians have increased from about 1.6 million tons in 1948 to approximately 5.4 million tons in 1966, or double the United States figure. But the United States fish industry has increasingly moved abroad to avoid domestic institutional barriers and to secure raw materials with which to supply its United States markets. In 1948 the United States imported 564,000 short tons of fish (round weight), which was 20 percent of its total needs in that year. In 1966 it imported 4.0 million short tons of fish (at a cost of more than $600 million), which made up 65.1 percent of its use in that year. Most of these imports came from the operation of United States

fish firms abroad using foreign fishermen out of foreign bases either in solely owned or joint ventures.

Unfortunately, the United States flag domestic fisherman has been deliberately suppressed in his activities in order to promote the exports of fish by poor allies to the United States, to promote social objectives in the United States, and to implement general governmental policy. Russia has not developed fishing technology herself but has purchased it abroad. In the current five-year plan, for instance, the Russian capital fund for improvement of its domestic fishery is $3.2 billion, a figure we would think of only in relation to space ventures, urban renewal, or the waging of war.

4. Promotion of aquaculture. There is now great enthusiasm in high quarters of the United States government to move agriculture to the sea in order to bring the domestic fishermen out of their neolithic state and to relieve the protein malnutrition of the world's starving millions by providing them with all the animal protein they can use. Japan is cited as the example we should seek to emulate. The fact is that aquaculture has been practiced for generations, in the United States as well as elsewhere. Products from aquaculture are high-cost delicacies: oysters, clams, mussels, abalones, shrimp, yellowtail, and flounder. Those people needing animal protein could not afford these delicacies by any stretch of the imagination or of their slender purses.

In Japan the raising of fish in ponds, which is done practically and on a considerable scale, is negative in its production of animal protein. The weight of cheap fish used in feeding the pond fish is greater than what results from the pond cultivation (Kasahara, personal communication, 1967). So long as there is much unused wild fish available in the ocean that can be harvested so cheaply, pond-raised fish cannot compete with it on a scale adequate to affect general human nutrition in a measurable manner. At present there are about 500,000 tons of fish meal available for sale in Peru at less than $100 per ton, or $.05 per pound, and it is 60 percent animal protein. Fish protein concentrate to human hygiene standards can be produced in the United States at $.25 per pound: it is 80 percent protein and gives a protein cost of $.31 per pound. Dried skim milk can be bought in the United States for $.15 per pound. It is 35 percent protein, giving a protein cost of $.45 per pound. I know of no fish culture anywhere in the world which produces fish or aquatic animals at a round weight cost of much less than $1.00 per pound, which would yield a minimum protein cost of $5 per pound.

SOME FISHERY PUBLIC POLICY MATTERS FOR CONSIDERATION

If the course of maximizing animal protein yield from the sea is taken, it will impinge sharply on several aspects of United States public policy and that of other governments. For example, the chief method used in conservation regulation in the great trawl fisheries of the northern hemisphere is keeping the mesh size of the trawl large enough so that the young flat fish, haddock, cod, or hake can escape through the trawl mesh and grow to a more desirable larger size. If, however, we wish to maximize the yield of animal protein from an area of continental shelf, we would need to have small-meshed nets.

We set great store by the excellent job we have done in conserving the Pacific fur seal. They eat about as much fish per year as the Pacific Coast fishermen of North America catch, but their prey are mostly deep-sea smelt and other such bathypelagic curiosities that the commercial fishermen do not now catch. If, however, we sought to maximize the yield of animal protein from the North Pacific, a choice would need to be made between fur seals and food, and the choice might be excruciating to professional conservationists.

The conservation of halibut has become a fetish in the Pacific Northwest, but the maximum sustainable yield of halibut possible from the northeast Pacific is about 35,000 tons per year. If a choice has to be made between 35,000 tons a year of halibut or 2 million tons of "trash fish" per year from the same grounds, the political anguish would be considerable.

Since the depths of the Depression, fishery conservation regulation in the Pacific Northwest has been aimed at reducing fishing effort by restricting efficiency. By this means not only were the fish stocks protected but also the inefficient small-boat fisherman was too, with obvious social and political objectives in mind. If one is to maximize the production of animal protein from a sea area, it will be necessary to catch large volumes of animals with maximum efficiency and at the lowest possible cost per ton of production. While the nation might well gain materially from such a decision, the short-term gain by the inefficient small-boat fisherman would likely be pretty negative.

We have pretty well abandoned a good many fisheries for clupeoids in western North America (anchovy in California, Oregon, and Washington; herring in Alaska) in order to provide "forage" for carnivorous fish that we like, such as salmon, bonito, and albacore. I know of no scientific evidence from anywhere in the ocean where a reduction in a forage organism has affected in a measurable way the abundance of a carnivorous fish that used it for food. I would doubt that such conditions could be brought about, because of the complexity of the web of life in the ocean, but this is one of the most cherished conservation gods we worship. Also, I would hate to be the politician who found it necessary to bring the matter to issue. But I can foresee the day when, if we took the course of maximizing the food yield from our coastal ocean area, we might find it desirable to weed out predators like salmon, halibut, albacore, and the like in the same manner we have weeded out the grizzly bear, timber wolves, and coyotes on the ranges where we wish to maximize the yield of sheep or beef cattle.

We abhor the thought of the federal government's running things, and particularly infringing upon states' rights. We feel that no state right is much more sacred than that of managing its fisheries. Consequently, our fisheries are managed under fifty separate, conflicting, archaic bodies. The further consequence is that foreign fishermen from Asia and Europe come several thousand miles from their home port, do their fishing a few miles from our home ports, and catch more fish off our coasts than we do. The reason is simply that we have protected our rich resources so assiduously from our own fishermen that other fishermen can afford to come thousands of miles because of the high catch per unit of effort they can get in our nearly virgin grounds. This is of no concern to us because we are a rich nation that has no foreign exchange problem and can afford to pay $600 million per year to foreigners for the fish we use. The nations of Europe and Asia, both friend and foe, must look to their foreign exchange balance and so they support their fishermen as we support our farmers and manufacturing industries.

How we are to cut this Gordian knot of state regulations which prevent our fisheries from developing and which protect every local whim and idiosyncrasy down to the county level in several states, other than by chopping it to bits with the heavy sword of turning fishery regulation over to the federal government, is not clear to me. Until we can unfetter the hands of our fishermen so they can go out a few miles from home port and compete on equal terms with fishermen who have come thousands of miles from home, I personally see little reason for plowing more millions of dollars into fishery development research in the United States. We now have more knowledge, technology, and market than we are permitted to use.

With those two sacrilegious statements, I close.

REFERENCES

Food and Agriculture Organization
1961   *Future Developments in the Production and Utilization of Fish Meal,*
       vol. 2. Yearbook of Fisheries Statistics, Nutritional Report Series.
Graham, H. W., and R. L. Edwards
1962   "The World Biomass of Marine Fishes." Pp. 3–8 in *Fish in Nutrition.*
       London: Fishing News (Books), Ltd.

Kasahara, H.
1966   "Food Production from the Ocean M.S.: Prospect of Fish Production from the Sea." Pp. 958–64 in *Proceedings, 7th International Congress of Nutrition, Hamburg.* New York: Pergamon Press.
Moiseev, P. A.
1965   "The Present State and Perspectives for the Development of the World Fisheries." Pp. 69–84 in *Seminar in Fishery Biology and Oceanography for Participants from Asia, Africa, the Pacific Area, the Mediterranean and Some European Countries.* Food and Agriculture Organization, EPTA Report No. 1937.
Schaefer, M. B.
1965   "The Potential Harvest of the Sea." *Transactions of the American Fisheries Society* 94, no. 2, pp. 123–28.
1968   "Economic and Social Needs for Marine Resources." Pp. 6–37 in John F. Brahtz, ed., *Ocean Engineering.* New York: John Wiley & Sons.

The Ocean Margins

I HAVE been asked to present a provocative discussion of the position of a marine state in acting on its marine problems and its marine opportunities. In attempting to do this, I will naturally concentrate on the state and the conditions that I know best: California and its waters, beaches, harbors, weather prediction, and fisheries opportunities. A provocative paper is likely to be either unsupportably optimistic and full of enthusiastic opportunities or, alternatively, deadly pessimistic, presenting a vast gathering array of towering and insurmountable problems. I will try to sketch the problems and opportunities as I now see them, noting both the hopeful and the troublesome in the proportions that I believe them to exist.

Before I discuss the specific matter of California and her marine concerns, I will place science, technology, and the public matrix in some of their present and evolving relationships. Perhaps the following three quotations can be used to define the extreme positions of science and politics. First, from the astonishing autoepitaph of the great naturalist, John Muir, written perhaps in 1905 and seemingly prescient of the Van Allen belts, solar winds, and the magnetic fields of space:

> I should hover about the beauty of our own good star. I should not go moping among the tombs, nor around the artificial desolation of men. I should study Nature's laws in all their crossings and unions; I should follow magnetic streams to their source and follow the shores of our magnetic oceans. I should go among the rays of the aurora, and follow them to their beginnings, and study their dealings and communions with other powers and expressions of matter. And I should go to the very center of our globe and read the whole splendid page from the beginning.

And from H. G. Wells's *Outline of History:* ". . . as scientific men tell us continually, and as [practical] men still refuse to learn, it is only when

John D. Isaacs is affiliated with the Scripps Institution of Oceanography, University of California, San Diego.

knowledge is sought for her own sake that she gives rich and unexpected gifts in any abundance to her servants."

I would like to talk about this matter of practical and basic research at length and about the contributions of marine biology to optics, astrophysics to petroleum geology, or cancer research to hedgehog hibernation, for example. But to continue, here is the final quotation, which is a statement made by a famous politician during the debate in Congress around 1805 over the desirability of the Lewis and Clark Expedition, which was granted federal funds in 1803: "What need have we to know of this thousands of miles of forsaken coast, peopled only by wild beasts and Indians. Why it will be a thousand years before we will have any need to go there!"

As I have said, I believe these quotations define the extremes of some portion of the scientific-public context, the untrammeled universal scope of science and the here-and-now concerns of the politician. I think that there is today some real reservation or even disenchantment with science. Some of this certainly stems from the lack of success or the relative lack of contribution of science to some of the vital problems of the moment, such as the Vietnamese war, crimes and violence in the streets of our cities, and many of the other apprehensions that are so strongly in the public mind today and, hence, in the minds of our politicians.

There is also a changing relationship of science and scientific discovery with one of our principal activities, industry, for in the United States we have arrived at a new phase in the effects of inputs on development. At one time the discovery of the salmon, or a new mineral lode, a deposit of fertilizer, or a new rich valley was an input that would greatly influence the inhabitants of the region. Today, with our complex institutions and our highly tuned economic and distribution systems, inputs are no longer strong. A striking example is the development of atomic energy in the form of nuclear reactors. This extremely bold, new discovery of a fundamentally new source of energy has had rather hard sledding. Not only has its development been constrained by public apprehensions over atomic energy and nuclear isotopes, but, more importantly, it must be fitted into an established economics of the power industry, which is not particularly sensitive to the cost of its raw power. You may, of course, argue that modern automation is a strong effect. I would submit, however, that high cost of help is the strong effect and that automation of a quite advanced type has been available for at least two hundred years—punch cards and all!

The burden of some of this disenchantment must fall on the scientists also. Personally, I have been increasingly concerned with the rather single-mindedness of much science. The last several hundred years of evolving science have been spent mainly with analysis—of taking things apart, studying their fundamental construction and laws, and attempting to build models that will replicate nature. This process has, of course, led to immensely important discoveries, but there continues to be a fixation on this sort of science as science of the highest caste. Perhaps this results from the development of the creative instinct in many scientists into what is almost tantamount to a theological concern. That is, in this age in which men are abandoning their gods as the agents of fundamental creation, it is perhaps unspokenly the scientist who believes he can fill this vacuum by assembling the appropriate equations and the appropriate relationships and show the universe to be recreated in his computers.

Although this is a great ambition indeed, it may be that, in the ultimate analysis of truth, this is not a possibility, and that the features and processes in the real world are generated in a system that is only very weakly constrained by basic laws, and that the events of the creation of this universe are in a large part the result of broad, nondeterministic selection from a very wide spectrum of possibilities. We may therefore truly need far greater attention to our integrative sciences. We have devoted little effort to the art or science of putting things together in a complex interdisciplinary system (which must even involve economics and the humanities). Perhaps we must look at the pragmatic descriptive approaches much more thoughtfully and confidently than we do today.

But, for the moment, I am more interested in the development of these aspects of new science, all reflected in the interdisciplinary development of oceanography, and the possibilities of looking at many of our other environmental problems in this fashion, by truly and confidently engaging the interest of specialists, experts, theorists, naturalists, and descriptivists. Perhaps we must start to approach the very complicated integrated problems with the same confidence with which chemists have traditionally pursued their problems, that is, on the pragmatic empirical basis, penetrating into basic principle only when advancing theory or opportunity permits, and formulating laws of interaction at the same empirical level as that of the objects of study. One need only consider some of the difficulties of the three-body problem in order to cast serious doubts that the science of chemistry could ever have been generated *de novo* from a knowledge of the fundamental laws of particles. At any rate, part of the disenchantment may perhaps stem from this inattention of scientists to the complex interdisciplinary problems of man and from the scientist's fixation in his demiurgic goals.

These changes appear in part in the breakdown of the traditional intermediaries between the research person and the user. These intermediaries have, in the past, provided the function of relating the long-term goals of research to the short-term demands. We see today a withdrawal of many of the traditional intermediaries, the regents or trustees of the great institutions of learning and research and the national sponsors of research from this traditional and, I think, highly effective role. This role provided an interpretive match and a perspective of time to the agents of the users, including, of course, state legislators, Congress, the military, foundations, or the great government research programs. These intermediaries also previously provided the feedback of interpreting need.

The scientist in research is thus increasingly in direct confrontation with the user and with the necessity of explaining or defending matters wherein he may not be very competent, such as the immediate, specific, and practical benefits to be derived from his research. The capability of a scientist to do this is not necessarily in a one-to-one relationship with his fundamental competence in research. Hence there is a selection of well-supported scientists toward the scientist salesman who, competent as he may be, does not necessarily represent the full range of competency of the scientific community. Nor is he likely to be able to "sell" the extreme importance of exploration, of untrammeled curiosity, of the more fundamental definition of human need, of aesthetics, or of the hypereconomic benefits of science.

This is supposed to be a provocative discussion; the very nature of this entire symposium is, as I see it, a thoughtful response to the breakdown of the intermediary system. Another of the many results of this developing conflict is that of the academic versus the commercial scientist. The commercial scientist, of course, is perhaps much better trained (and surely far better motivated) to sell ideas for immediate and perhaps demonstrable benefit. One result of this is a proliferating number of rather shallow advocates of certain single-minded developments in all sorts of fields. Many of them are very important fields, indeed; but most of these developments, even in combination, do not suffice to solve the fundamental problem.

I should perhaps make sure that you understand my points. I strongly believe it is necessary for many scientists to take a deep interest in practical problems. H. G. Wells has said: "We have still to ensure that a man of learning shall be none the less a man of affairs, and that all that can be thought and known is kept plainly, honestly, and easily available to the ordinary men and women who are the substance of mankind." But those scientists who are incapable of being men of affairs make no less of a contribution, and we will immediately cut ourselves off from vital understanding if we do not also support the retiring specialist. There is nothing incompatible in arguing for the full spectrum, although there may superficially appear to be an inconsistency. It is difficult to formulate convincing arguments for both motherhood and chastity in the same paragraph.

This preliminary discussion leads almost directly into some of the ocean margin problems of the state of California, with which I am quite well acquainted. I should now leave the defense of the above ideas, the elaboration of some of the points, and a certain amount of retraction to later discussion. Many of the general points I have already made will more specifically emerge from the discussion of California's marine problems.

The problems of California that I will scan include those of water—including the matter of desalination—beaches and beach erosion (curiously related to the water problem), and the problem of weather prediction, very clearly interrelated with water. I will mention recreational harbors and I will deal somewhat with some of the problems of California's marine fisheries. I will also say something about the organization of the state for the guidance and the solution of her marine problems and, in passing, mention some of the large research programs on marine resources conducted at the Scripps Institution of Oceanography, University of California.

The water problem is certainly one of the strangest, least understood, and difficult problems society faces in arid zones. I could readily present you with several chapters on this peculiar subject, but I will have time only to cover the part that relates to California's technological and institutional competencies to manage her water supplies. It can be easily shown that the only real problem of water in any technologically competent nation is that of agricultural water. Table 1 will demonstrate the rather trivial economic

TABLE 1
APPROXIMATE ANNUAL WATER COSTS
(per capita)

Type of Water	Irrigation [a]	Drinking [b]	Domestic and Municipal Use [c]
Shallow pumps, ground water	$ 6.00	$ 0.005	$ 0.30
Forage crop irrigation water	15.00	0.01	0.75
Wholesale irrigation water (S. Calif.)	36.00	0.03	1.80
City domestic water	400.00	0.30	20.00
Desalinated sea water (attainable low)	600.00	0.50	30.00
Desalinated sea water (navy cost)	1,500.00	1.10	75.00
Bottled drinking water	240,000.00	180.00	12,000.00

[a] 3 acre feet per year—enough to raise food for one person for one year.
[b] 1 ton per year—enough to supply one person's needs for drinking and cooking.
[c] ⅛ acre foot per year—for all municipal uses.

problems of supplying cities, industry, and households with all of their water needs. Without going much further into nonagricultural use of water, it is quite clear that the production of water by desalination for any municipal, industrial, and domestic use in any advanced country is quite within the competency of the present technology at an annual cost perhaps something on the order of thirty or forty dollars per capita, even from present desalination stills that have existed for the past twenty or thirty years.

The cost of water for agriculture is quite a different thing. It is foreseeably impossible to supply any significant amount of water from desalination for agriculture other than of highly subsidized crops or of crops for luxury consumption. California, for all of its concern with water and the high prices that it must pay for water from outside sources, manages to use much of its local water supplies in extremely ineffective ways. It has been pointed out that the subsidies to arid-region agriculture, including the subsidies to irrigation water, the overcharge of domestic and municipal users, the tax-free character of agricultural water-bond developments, and the direct subsidy of the crops, are such that many agricultural activities in arid regions are so highly subsidized that the crop is paid for by the public before it is removed from the ground. I think that this is an overstatement, but it is true that California will pay upward of seventy or eighty dollars an acre foot for new sources of water, while it still supplies water for raising low-grade agricultural fodder crops at a price of as little as one dollar per acre foot, or less than one-tenth of a cent a ton—surely the cheapest material on earth.

I spoke earlier of the shallow way in which many practical problems have been studied and of the resultant ineffectiveness of the solutions. Figure 30 shows one of the characteristic tools of the single-minded advocate. This map, showing water shortages in the United States, was used as part of the evidence for the initial granting of funds for a federal organization to conduct desalination development. Most of the shortages are clustered around the Great Lakes, along the Mississippi River, along the Columbia River,

Figure 30. Areas reporting water shortages in 1957

and in other such places where it is difficult to conceive that desalination can provide any reasonable solution to the water shortages. Even in the more arid regions coast to coast it can be rather easily shown that because of pumping costs, were the oceans fresh water, it would still solve very few of the water shortages.

My point here is that the problem of water has first been too shallowly defined and then turned over to a group to advocate desalination. It is rather difficult to ascertain how this particular direction of approach was adopted, but it certainly was not from any rational analysis of the real needs for water or of the problems that could be solved by desalination. Remembering that the problem is entirely one of food and agriculture, the definition of the problem then, of course, is not one of the availability of water, but of the raising of food. This definition vastly broadens the problem and increases the number of reasonable solutions by orders of magnitude. On a planetary basis, we have, for example, the alternative to learn to cultivate the wet lands; but we know almost nothing about it and what little of it we do know is inadequate.

In another direction, many higher flowering halophytic plants are capable of living in sea water or water of even greater salinity. Yet there are essentially no breeding experiments that I can trace which attempt to introduce this genetic knowledge of the natural desalination of sea water by halophytes into our crop plants so that they can resist the ubiquitous salination of soils, or even be raised by sea-water irrigation. Perhaps this tiny molecule of DNA that can "teach" crop plants how to desalinate sea water with solar energy is the most precious of our marine resources.

However, there have been several recent break-throughs in the direct culture of ordinary crop plants in sea water. Some years ago it was discovered in Israel that crop plants could be raised in gravelly soil that was periodically irrigated with saline water. The exact manner in which the plants managed to survive and grow well under these circumstances is unclear, but higher plants that are ordinarily not salt-tolerant have been grown hydroponically in agitated sea water.

A possible theory now ties many experiences together. One difference between a salt-tolerant plant or halophyte and an ordinary plant may be that the halophyte has a particular capability of supplying oxygen to its roots to carry out the metabolic job of desalinating sea water. Thus ordinary plants when grown in saline soils do not die from the salinity directly but rather from root asphyxiation because they are unable to supply sufficient oxygen for the osmotic job. An alternative theory, of course, involves the transport of concentrated salt away from the immediate area of the rootlets. Clearly, the redefinition of the *water* problem as the *food* problem opens up a multitude of very promising paths, quite unlike the constraint of narrowing it to the single solution, desalination.

With this brief treatment of the water problem let me pass on then to the related problem of beaches and beach erosion in California. The usual problems resulting from intervention into the coastwise transport of sand are quite well understood around the world. Harbor works are almost always undertaken with the expectation that the sand accumulated as a result of interventions must eventually be moved.

In another respect, however, California is in a particular predicament. The normal balance of sand on the beaches of the coast of California is the result of a transport of sand down river washes onto the coast, followed by its littoral transport. It then accumulates in certain regions and cascades downslope into the deep oceans through submarine valleys, or in a few locations it is blown ashore into dunes. The stability, in fact the existence, of beaches depends on this transport phenomenon and hence on the inputs. However, flood control projects in streams have interrupted the flow of sand *to* the beaches, but the flow away from the beaches and down the submarine canyons and to the other sinks remains uninterrupted. California is thus faced with an evolving problem of a long-term sand starvation along her important beaches. Her ultimate choices are either to remove the material accumulated in the flood-control reservoirs and reinitiate the transport onto the beaches or to return the sand to the coast from the deep continental slopes and underwater fans.

This problem can be put off for some time by the dredging of accumulations of sand in the dry or very shallow lagoon valleys along the California coast for the construction of small boat harbors. These harbor developments are a large part of California's future, and the development of Mission Bay near San Diego demonstrates the long-term possibilities of California's greatly increasing its shoreline. I believe these developments are aesthetically pleasing and delay the ultimate saturation of California's beaches for recreation and building. There have been many surveys of these possible sites for small boat harbors extending back some thirty years or more, and California is now quite conscious of the necessity to maintain these important sites uncommitted to uses incompatible with their ultimate development as harbors, particularly in southern California.

I will devote the rest of this discussion to other problems and opportunities of California in two interrelated major areas: the fisheries and the prediction of weather. The fisheries problem provides a particularly illuminating and pertinent example for this symposium, for not only do great opportunities exist for the development of important fisheries in the waters off the California coast but California displays an inability to act effectively on these resources. The case thus points up the profound influence of the institutional barriers that so commonly exist in many of our activities in the United States today.

Let me first sketch a bit the history of the major fisheries of California.

The major commercial fisheries of California have been of two types: the first is a purse-seine fishery for the sardine, which reached its peak in the 1930s with some half million tons of sardines annually extracted from the waters of the Pacific; and the second is a luxury fishery for tuna, which developed in California and has more recently expanded widely around the world, introducing new methodology and competing quite successfully with foreign fisheries.

The sardine fishery began its growth in the 1920s, reached its peak in the 1930s, and declined during the 1940s, virtually disappearing by 1950. The reasons and nature of its disappearance have been the subject of intensive controversy and also of scientific investigation by the University of California's Scripps Institution of Oceanography, the California State Department of Fish and Game, and the Bureau of Commercial Fisheries. At present there is perhaps no fish in the world (particularly an almost nonexistent one) that has been so intensely studied as the sardine, nor an oceanic region of the world that is so well known as the California Current, as a result of some twenty-four years of study.

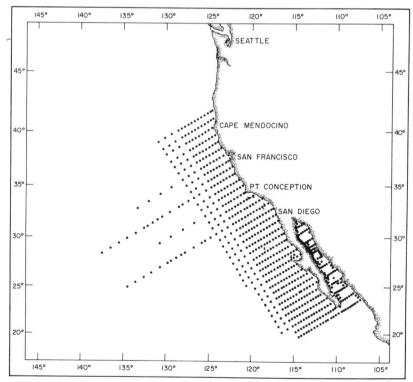

Figure 31. Array of sampling stations used for survey of the sardine population (CALCOFI Basic Station Plan)

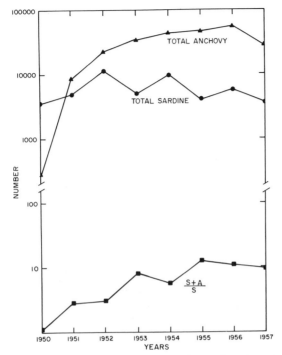

Figure 32. Populations of sardine and anchovy larvae based on annual surveys, 1950–57

Figure 31 shows the area, nearly a third of the size of the United States, that was covered with almost monthly surveys for a period of twelve years up to 1960 and has subsequently been surveyed at less frequent intervals. The final decrease in the population of sardines, following a short resurgence, is well shown in Figure 32 which totals the larvae of the fish found annually throughout the range. The graph also shows the concomitant increase in a competing species, the anchovy. It is now apparent that the anchovy exists in the California Current region in at least as great a number as the sardine did at its greatest abundance. These larvae enumerations are a very effective measurement of the adult population, as the fecundity of the adults is quite well understood. Also, the larvae are rather easily captured quantitatively with simple plankton nets and it thus is possible to survey the entire habitat of the creature.

Much public controversy was stimulated by the allegation that an unrestrained fishery on the sardine had destroyed the stocks of this fish and allowed the anchovy to take over its normal habitat. It is quite strongly indicated, however, that the anchovy is the normal inhabitant of the region and that the sardine fishery was initiated during an unusual period in which the sardine was in a temporary abundance.

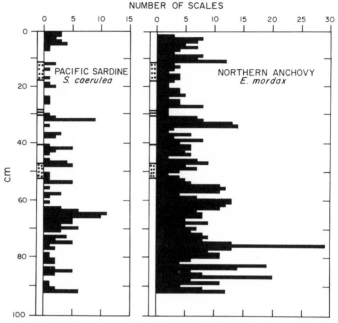

Figure 33. Abundance of sardine and anchovy scales in centimeter sections of bottom sediments, Santa Barbara, California

Figure 33 demonstrates something of the knowledge that has been obtained from some rather remarkably preserved varved sediments that are found in the deep Santa Barbara Basin off southern California. The individual layers of these sediments represent the annual deposits of organic debris and sediments from the overlying layers of water of the California Current System. As you can see, anchovy scales are in abundance at all depths. The entire time period covered by this chart is around two thousand years, during which time there is never a period in which the anchovy scales are absent, except in the very uppermost layers that represent perhaps the last forty years or so. At the same time, the deposits of sardine scales are very irregular and there are only two periods of abundance, the very recent period of perhaps seventy years and a similar period of abundance some thousand years ago. Both the larvae and the scale data indicate that the hake, a pelagic fish in that part of the ocean, has been in continuous high abundance, varying somewhat similarly to the anchovy.

We have thus been able not only to establish a very intimate history of the decrease of the sardine in the California Current System, and its replacement by the anchovy, but have been able to show that the anchovy and hake are probably the stable component of the environment over very long periods of time. Yet only with great difficulty has the California legislature been able to permit the initiation of a small fishery on the anchovy. This difficulty has ensued almost entirely from the apprehensions of the sport fishermen and their allegation that it has been the commercial fishermen who have been responsible for the decrease and disappearance of the sardine. There is some real basis for this allegation, for it is quite clear that the fishery developed essentially without restraint and it was only after the onset of its decline that any very serious or complete scientific investigations were begun.

The sport fisherman is, of course, able to think up many objections to a new purse-seine fishery; but in the ultimate analysis, I believe that all the objections have but a single root—a lack of faith in the ability of the state of California to control a fishery on the scientific evidence. A prima facie case of California's inability to act on the scientific evidence can today be shown in the slow decrease of the population of Pacific mackerel, the California Fish and Game Department's urging that controls be placed upon its capture, and the state's failure to do so. There seems to be no question that this species is seriously overfished, and, although it is not presently of great economic importance, the case nevertheless clearly demonstrates that the sport fisherman's apprehension can be well founded.

To return to the anchovy, we thus have the strange anomaly of an abundant supply of a pelagic fish of considerable economic significance, a foreign harvesting of this fish, and a domestic inability to solve, not the scientific or the technological problems, but those institutional problems that prevent the state's benefiting from this economic opportunity.

You can see from this example, as in the case of water, that it is quite possible for the scientist to devote his serious thoughts and efforts to understanding a matter of potential commercial benefit, and that it is also possible for this thought and effort to be misused. The scientist is frequently the wrong person to approach the problem, since the fundamental barriers are often not the result of a paucity of scientific information but rather are of a social or political nature. We see more and more frequently research erected as a façade to obscure society's inability to solve some institutional problem, and both society and scientists emerging with empty answers, or none at all, to shallowly defined problems.

These social and political barriers also develop from a lack of appreciation of the hypereconomic benefits of research and of developing such resources as fisheries. For example, one can legitimately view the development of a pelagic fishery not only as a source of immediate income but also as a source of experience in fisheries science and technology. Economic gain subsequently comes from the exportation of this scientific and technological know-how by industrialists, entrepreneurs, fishermen, and scientists, as they are enabled to recognize opportunities for fisheries development elsewhere in the world. The larval technology, for example, permits discovery of stocks of commercially valuable fish without the complexity and expense of mounting an exploratory fishery, with its uncertainties of the magnitude of the stocks and of the availability of the fish to the gear.

The work on the fish scales in the sediments represents a potential, and sophisticated, way of guiding a fishery and of achieving a degree of understanding of the fishery at its initiation that cannot be achieved in a century of fishing and conventional scientific investigation on the stock itself. One might think that this argument on the potential economic benefits of such experimental fisheries is farfetched or unlikely. In California, however, we have a demonstration of this exact mechanism in the tuna fishery. The tuna fishery developed initially off California and was highly experimental in the technology of catching and processing the fish and in appraising customer acceptability of the product. It developed its technology almost without restraint by operating in distant waters, out of the jurisdiction of the state of California. Since then it has been able to explore the most effective methods of catching, preservation, and product handling. Opportunities have developed around the world for United States entrepreneurs. Profitable canneries, all involving California and United States companies and investment, have been established in distant lands. California scientists guided research and fisheries development in these other regions. This certainly provides an astonishing contrast to the manner in which the potential anchovy fishery has been handled, where the sport and commercial fishermen squabble like children on the beach while the tidal wave of foreign exploitation poises to sweep their toys away.

Not unexpectedly, the entire matter of fisheries research in California has led to a number of wide-ranging investigations of the nature of weather variation, both oceanic and atmospheric. It has long been known that the success, the nature, and the composition of the oceanic animal populations have been greatly altered during changes in water temperature. Along the California coast this has been particularly conspicuous during periods such as the 1860s, when tropical organisms lived in Monterey and visited as far north as Puget Sound. Changes in ocean conditions are associated with fluctuations in the success of fish spawnings, the type of copepods present, and other alterations in the marine environment. However, only since the initiation of these large investigating programs has it become clear that the major changes in the ocean climate and in the atmospheric weather of the west coast are not mere local changes, but parts of large-scale fluctuations in the ocean-atmosphere system involving the entire North Pacific, or more probably the entire hemisphere—or the entire planet.

Figure 34 shows two-year and ten-year running means of the air temperatures at San Diego over the last one hundred years. Here can be seen rather rapid short-term variations involving three- to five-year periods and a long-term secular change with a great cool period in the 1880s and 1900s. The

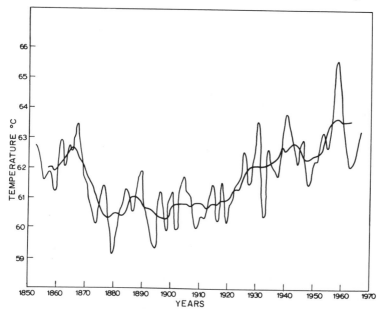

Figure 34. Two-year and ten-year mean annual air temperatures at San Diego, 1850–1965

mean ocean and air temperatures are closely related in marine locations such as San Diego, with some recent exceptions, and we can look upon the air-temperature record as a historical record of sea-surface temperatures. Some studies have shown that short, intense fluctuations in climates are related in places as distant as Peru and Japan. We now believe that this relationship is probably an expression of the very large dimensions of ocean-atmosphere change, rather than the result of some mysterious teleconnection, as has been implicated.

The most conspicuous recent change in the North Pacific conditions started about 1956. The anomalous sea-surface temperatures in the North Pacific during the period 1956–57 are shown in Figure 35. Higher than normal temperatures appear in stripes and those colder than normal are shown in dots. These changes have been associated with and probably caused by changes in the atmospheric circulation, but the atmospheric circulation was again modified by the abnormal ocean conditions.

The changes in the marine organisms and their distribution as a result of shifts in the ocean conditions provide an entree into extending knowledge of the ocean atmosphere climate of the past. I have already spoken of the record of the fish scales in the varved sediments of the Santa Barbara Basin. Of course, there is other organic debris recorded there like enciphered pages of oceanic and meteorological history. Part of this reading comes from the

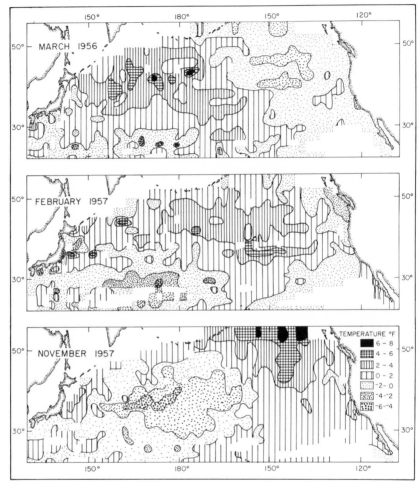

Figure 35. Deviation of sea surface temperatures in March 1956, February and November 1957, from 1947–58 mean temperatures

distribution and occurrence in the sediment of such creatures as the little pteropod, *Limacina helicina*. Figure 36 shows the distribution of this pelagic pteropod, which inhabits the waters of the Gulf of Alaska and the upwelling regions of the California Current. Its presence off southern California is thus a qualitative measure of the strength of the California Current.

In Figure 37 is shown the distribution of the shells of this microscopic pteropod in the sediments of the Santa Barbara Basin, plotted against years from about A.D. 900 to 1850. Plotted also are ten-year discrete means of the air temperatures at San Diego as were shown in Figure 34. You will see

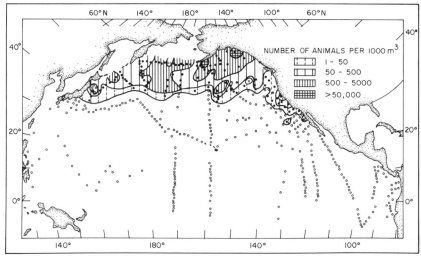

Figure 36. Abundance of *Limacina helicina* in the North Pacific Ocean

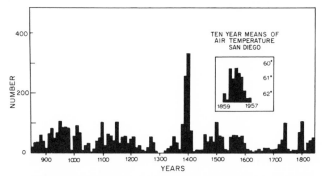

Figure 37. Abundance of shells of *Limacina helicina* for ten-year intervals of sediment in the Santa Barbara basin and ten-year means of air temperature at San Diego

that the two records have much the same character. Other organisms will give us other answers, such as the contribution of subtropical waters, the encroachment of Central Pacific waters, countercurrent flow, and also the atmospheric conditions that give rise to such changes. Altogether we hope to be able to assemble an imposing history of the nature of climatic fluctuations with a very high time resolution—that is, with a time resolution interesting to the events in a single human life, but also extending into the last several thousand years. Similar sediments as these in the Santa Barbara Basin exist in some basins of Baja California, off the Central and South American coasts, and also in Sannich Inlet at Vancouver Island. We have started an investigation of these Sannich Inlet sediments, and it appears that they are very well stratified, presumably with annual varves. The organic debris is well preserved and appears to include tests of pelagic organisms, but fish scales do not seem to be conspicuous.

The cases that I have discussed briefly here are only a scattering of the marine investigations of interest to the state of California and are mainly ones conducted by the Scripps Institution of Oceanography under its Marine Life Research Program, with the Bureau of Commercial Fisheries Laboratory in La Jolla and with the State Fish and Game Laboratories in Terminal Island. There are, of course, many other such investigations involved with coastal structures, offshore oil exploration and development, pollution, atomic waste disposal, estuarine problems, sport fisheries, and heat rejection from power plants.

The mechanisms by which these researches and the consequent advice are conveyed to the state government and to the legislature are, of course, studies and reports on particular matters, which may be carried out and reported by the Resources Agency, or which may be called for by the legislature (or carried out by the legislature and its committees). There also are a number of standing groups that are advisory to the state. For some years the Marine Research Committee, part of the California Fish and Game Department, has been supporting research within some of the state laboratories, the University of California, Stanford, and the California Academy of Sciences, mainly on the problems and possibilities of pelagic fish. I believe that this organization has been a very helpful one, although its attitudes have been principally those of commercial advocates and hence its advice has been held in some distrust by the sport fishermen. A much more comprehensive area has been covered by GACOR (Governor's Advisory Commission on Ocean Resources). This commission reports directly to the governor and advises studies and controls related to the marine environment.

So far I am not highly optimistic for California's purposeful handling of her marine opportunities. The most effective developments have been those that have escaped from state control. Although the oceans are certainly either California's boundary or its open road to opportunity, depending on how effectively she explores and understands her opportunities, it is quite possible that these opportunities will mainly be reaped by other political entities, foreign or domestic, who have a more farsighted viewpoint, are freer to act, or for whom these possibilities represent broader, stronger, and more urgent openings than they do for the state of California.

To summarize my principal points, I have first submitted that there is some substantial disenchantment with science, which stems from a lack of understanding on the part of the scientist as well as that of the user. We need both more intensive and more encompassing scientific approaches, especially in environmental science, and oceanography, by virtue of its multidisciplinary nature, can point the way to these new approaches. Definitions of problems are now often shallow, causing disappointment to the scientist and society alike, and we have strong hypereconomic reasons to define our practical problems in very broad terms. I have also submitted that the state of California often does not do well in acting on marine opportunities, mainly as a result of institutional barriers within the state, and therefore the most effective use of marine resources may well develop elsewhere.

Energy from the Oceans

RICHARD A. GEYER

THE title of this paper, "Energy from the Oceans," is quite ambitious. If all the aspects pertinent to this broad subject were to be discussed in detail, they would themselves constitute the basis for an entire seminar series. Accordingly, the subjects in this presentation will be more limited in scope. I shall concentrate on marine petroleum deposits rather than attempt to cover in detail a number of other marine energy sources of possibly great future potential.

For purposes of comparison and to provide a broader framework for discussion, competitive energy sources, both on land and at sea, will be considered briefly. These sources include tar sand, nuclear energy, coal, water power, and oil on land; and tides, surface waves, internal waves, and microbial sources in the sea. Also, the state of scientific and technical knowledge of offshore oil exploration and production will be reviewed.

In conclusion, I shall attempt to highlight pertinent interdependent facets of the total problem, including economic, legal, and political aspects, especially as they might be relevant to public policy. It should be emphasized that the statements made in this discussion are those of an individual and are not to be construed as reflecting the ideas, opinions, or policies of any group with whom I am currently affiliated, or have been in the past.

A BRIEF HISTORY OF OFFSHORE DEVELOPMENT

It is generally believed that the first commercial offshore production came at the turn of this century from the Summerland Field in California. Wells were drilled either directionally seaward from the beach or on wooden piers. The first major offshore oil production was in Lake Maracaibo, Venezuela, starting in the 1930s. Again, because of the shallow depth, platforms were erected on pilings driven into the lake bottom. Late in the 1930s, off-

Richard A. Geyer is professor of oceanography at Texas A & M University.

shore oil production began in the Gulf of Mexico with the discovery of the High Island Field off Texas, followed a couple years later by production from the Creole Field in Louisiana. Subsequent increased activity in the Louisiana marshes was curtailed during World War II. Since then, the search for and production of oil in the continental shelf has increased at a tremendous rate, first on the shelf of the Gulf of Mexico, later off the California coast, and finally on a world-wide basis. In 1946 the first commercial oil was produced from wells out of sight of land; now production is coming from sites more than seventy miles from the shore and in water depths of about three hundred feet.

It is estimated that at this time the oil industry has invested about $10 billion in world-wide offshore oil activities. The return on the dollar from offshore activities is still far behind the initial investments; but investment and returns are continuing at an increased rate. For example, the value in 1966 of oil and gas produced from the United States continental shelf was about $1.1 billion as compared with about $750 million in 1964. Similarly, the investment, just in leases and royalties paid to the federal government over the last eight years, was about $1.5 billion.

Turning now to the Pacific Northwest, exploration and production activities for oil on the continental shelf are a more recent endeavor, beginning in the early 1960s and intensified following passage of the Oregon Submerged Land Act in 1965. It has been estimated that about $77 million were spent in these activities off the coasts of Oregon and Washington. In addition to offshore exploration costs, lease bonuses and yearly rentals in the five-year period between 1961 and 1966 amounted to about $43 million. From the standpoint of financial benefits accruing to this area from offshore oil exploration, an additional $30 million were spent by seismic crews and drilling contractors working out of Oregon ports. In spite of these large and diversified expenditures, as of the moment no commercial oil has been produced. Nevertheless, sufficient information has been obtained from the drilling of the wildcat wells to keep this offshore geologic province in a category of having future potential, since a great thickness of sedimentary rocks has been proven to exist. Also, some oil has been found in two or three wells off the Washington coast.

The foregoing financial statistics naturally lead to the question: Why has the oil industry made such a major effort in both domestic and international operations in continental shelf areas? The answer has scientific, economic, and political implications. At the time when the domestic oil industry began to explore seriously the United States continental shelves for oil, similar exploration on land had been going on for more than fifty years. Tremendous reserves had been found during this time, many of them in very large fields; but unusually large forecasts were projected for future oil requirements. It became apparent that the best chance for success in finding these major reserves lay in hitherto unexplored areas where geologic conditions were comparable to those where large reserves had been found on land.

The continental shelf areas fulfilled these prerequisites. Concurrently, scientific and technological advances in oil exploration and production had reached the point where the search for and production of oil in the marine environment could be technically feasible and economically justifiable, at least on a long-term basis. In additiion, whatever reserves could be developed on the United States shelf would also help to maintain the domestic self-sufficiency of the nation. This became a factor because at about this same time oil companies had begun to allocate an increasingly large proportion of their exploration and production budgets to search for oil on land outside of the United States.

After several years had elapsed, it became evident that in addition to scientific and economic factors affecting oil exploration in foreign countries, the question of political stability became an overriding consideration in the decision to explore, or not to explore, in certain foreign areas. Furthermore, if a large oil deposit were found inland in a foreign country, there was frequently the problem of constructing a pipeline to bring the oil to sea for export by tanker to other parts of the world. Upon occasion the pipeline would have to cross one or more additional countries of questionable political stability, posing further political problems. Under these circumstances, it is not difficult to see that exploration on continental shelves of foreign countries has many important advantages. These considerations, again coupled with improvements in marine exploration techniques, resulted in increasing emphasis on development of oil reserves in foreign continental shelf areas. This activity continues at an accelerated rate even to this day.

DISCOVERY METHODS

Having discussed various reasons on both the domestic and international scene for turning to the sea for oil, it is in order to present briefly some of the scientific and technological factors involved in discovering methods that made exploitation of the undersea resources feasible. The successful and efficient development of a large-scale marine exploration program on the continental shelf falls into two phases: first, the use of reconnaissance techniques to localize areas of interest; and second, the development of more specific geological and geophysical information in these areas, using detailed exploration methods. In an area where there are underwater outcrops, scuba diving has been used extensively for reconnaissance information. Then actual cores in areas of interest are taken to obtain additional structural and stratigraphic data. This second method must be used exclusively in this phase of operation where outcrops are obscured by recent sediments.

Sometimes concurrently with this phase of exploration and sometimes even preceding it, aerial magnetometer surveys are made to delineate major areas of subsurface structural interest. The next reconnaissance tool generally used is marine gravity surveys, which are made either by placing the instrument at the bottom or, in more recent developments, by using an instrument capable of obtaining the necessary information continuously aboard ship. These results are studied and synthesized and are used then as a guide to define areas where first reconnaissance and then detailed seismic reflection surveys are to be conducted. Direct information regarding the subsurface structural conditions that might be conducive to the accumulation of oil and gas, in commercial quantities, are generally best obtained by the use of the seismic methods. The use of combined gravity and magnetic surveys, when interpreted properly, can lead to solving major geologic structural problems indirectly; but only rarely are drilling locations selected without the use of detailed seismic surveys.

PRODUCTION METHODS

The evolution of drilling methods for producing oil found by the exploration program described above is interesting. As the search for oil progressed from shore and beach areas to shallow water, drilling platforms were merely modifications of those used in marsh or lake areas. As the search for oil continued into deeper waters, drilling barges were used more extensively, and producing wells were completed on fixed platforms. As the search continued into even deeper waters, of the order of several hundred feet, the complex problems of underwater completion became increasingly important to solve because of economic considerations. As the drilling for oil continues into water depths of from six hundred to one thousand feet, underwater drilling and completion techniques must be relied upon entirely to solve not only economic but environmental problems as well.

It is evident even from this brief consideration that the successful solution of these problems requires a sophisticated knowledge of the oceanic environment around the areas in which these operations are being conducted. This in turn requires greatly improved ocean engineering techniques, as well as major advances in basic oceanographic data acquisition and research activities. In addition, once oil is found, one still has the problem of efficient transportation and economical storage facilities on land for further transportation to markets. The best solution is the construction of submarine pipelines, which also requires a broad knowledge of oceanographic factors, including marine soil mechanics. Major underwater storage depots have been suggested as production is achieved at increasing distances from shore and as tankers are filled directly from these points. Recent suggestions include the use of giant inflated rubber balloons or of underground nuclear blasts to carve out major subbottom storage caverns.

The advent of increasing exploration and production activities in Alaska, the Bering Sea, and Hudson Bay requires solving even more complex environmental problems. One suggested solution is the employment of submarine tankers to transport the oil on a year-round basis from the wells to market. The rigors of the Alaskan winter, even in the Cook Inlet area, are such that studies are now underway to determine whether underwater drilling as well as completion methods, using the man-in-the-sea approach, would be technologically and economically feasible. The same would apply to the previously mentioned arctic areas. A tremendous expansion in our research efforts in both oceanography and ocean engineering in the next few years is therefore imperative if technological progress is to keep up with the ever-expanding consumption requirements. In this regard, let us examine briefly some of the specific technical problems that exist in the production of off-shore oil.

LEGAL AND ECONOMIC ASPECTS

The legal and economic aspects of deriving energy from the oceans, especially in the form of petroleum, comprise some of the more controversial, as well as complex, problems in the search for and production of offshore oil, particularly if it is to be accomplished within the framework of a free enterprise system. Included in this category is the multi-user problem, which will become more complex as man looks to the sea for an ever-increasing variety of uses. This complexity in fact is compounded, if we may digress for a moment, by the rapidly expanding use of petroleum other than as an energy source. Its value as a raw material for petrochemicals is well established. But more recently, as a result of research by some oil companies, crude oil is beginning to hold great promise as an inexpensive source of protein. This potential in a sense gives added impetus to the search for and proper utilization of marine oil reserves.

At the moment, at least sixty countries are involved in varying degrees in offshore oil exploration and/or development, and offshore oil is produced within the territorial limits of at least eighteen countries. In fact, about five million barrels per day of offshore oil is presently being produced, which accounts for about 16 percent of the production from free-world reserves. In addition, it has been estimated that a quarter of the oil and gas requirements of the free world will be obtained during the next generation from offshore areas.

The rapid expansion of the offshore oil industry creates an urgent need for well-established and recognized legal procedures to establish claims for ownership of offshore oil and gas leases. Such procedures should be the concern of cognizant groups on state, national, and international levels. For some time in this country the coastal states have been battling the federal government over the right to lease offshore lands for oil. This controversy includes such questions as geographic criteria for determining the boundaries between state and federal ownership. But once these boundaries have been adjudicated, there remains a very practical problem of their absolute determination at sea, where there are few, if any, landmarks for this purpose. Consequently, scientific and engineering research is being conducted to develop suitable instruments to insure that the lease boundaries shown on a map can be accurately defined in space on the continental shelf. Nevertheless, the farther we proceed from the shore, accuracy becomes more difficult to attain; and accuracy is crucial to those who are spending literally tens of millions of dollars to produce the oil in a certain area.

The problem is complicated further where a definition involves not only one or more states of a single nation, but where accurate boundaries must be defined, after the criteria for their delineation has been determined and agreed upon, by a number of nations bordering an offshore oil province. One of the best examples is in the current development of oil and gas reserves in the North Sea.

Certain criteria were promulgated and subsequently adopted by many of the coastal nations during the 1958 Geneva Convention on the Law of the Sea. Included in these deliberations were definitions of the continental shelf and of certain geographic criteria establishing the legal rights of these nations. One of these articles applies to the geographic case where the same continental shelf is adjacent to countries which encircle it. Under these circumstances, the convention states that the boundary of the continental shelf should be determined by agreement; in the absence of other agreements, the boundary is generally considered to be the median line, equidistant from the nearest base line point of the territorial sea.

When the minimum ratification quota was achieved by Great Britain's acceptance of this convention in June 1964, the median line principle could then serve as the criterion for negotiations to divide the bottom of the North Sea among the contiguous nations. The entire North Sea is less than two hundred meters deep, except for a deep narrow trough off the Norwegian coast. Its continental shelf covers about half a million square kilometers, and the acceptance of the median line principle meant that the bordering countries were allotted the following areas, in square kilometers: Great Britain, 240,000; Norway, 131,000; The Netherlands, 162,000; Denmark, 54,000; West Germany, 22,000; and Belgium, 4,000. Because of the concave coastline of West Germany, this principle resulted in that country's receiving only a small section of the North Sea continental shelf. West Germany submitted a claim, resolved by the International Court of Justice, in an effort to obtain a larger portion.

Using the Continental Shelf Act of 1964 as a basis for its jurisdiction, Great Britain established leasing regulations for offshore petroleum production, with the responsibility given to the British Ministry of Power. It divided the British portion of the North Sea into 1,300 blocks containing 100 square miles each. Thirty-one applications had been received for 400 of these blocks by the fall of 1964. An extensive geophysical survey program had been undertaken earlier, between 1959 and 1963, by major oil companies, in anticipation of future lease grants. Based on these exploratory efforts, a number of drilling sites were selected by the fall of 1965. Subsequently, the first commercial oil was discovered, followed a few months later by a commercial gas discovery.

Various countries have different types of leasing agreements. For example, Denmark granted an exclusive fifty-year license to a single combine. In contrast, Great Britain has two types of licenses: one, a nonexclusive exploration license for exploratory work only; and two, the production license, permitting oil companies to drill for oil and gas in the allocated blocks described above.

Although the legal problems of jurisdiction seem to be solved in general in the North Sea, those of an oceanographic and meteorological nature affecting the production and submarine pipeline transportation of oil and gas in this area are quite formidable. The combination of shallow water and long periods of prevailing high winds results in very rough seas over much of the year which hamper offshore operations. In addition, the multi-user situation in this area adds further complications. To avoid interference with shipping and fishing activities, it seemed logical to plan to bury pipelines at depths of about five to ten feet under the bottom of the sea. However, this procedure becomes complicated sometimes in areas where a mixture of clay and boulders exists in an otherwise sandy sea bottom. This again emphasizes the need for increased knowledge of soil mechanics properties of marine sediments, and the development of methods for rapidly and accurately mapping changes in sea bottom composition where pipelines are to be laid or drilling platforms are to be established.

Multi-user problems are becoming increasingly acute in areas of high population density combined with high standards of living and advanced technology. It is not difficult to see where conflicts could arise in coastal and offshore areas between various groups having different objectives for the use of the same waters.

In addition to conflicts arising between different industrial and commercial users, there are also, particularly in many parts of the United States, the overriding requirements for the use of coastal waters for recreation and the aesthetic enjoyment of nature. This attitude is especially prevalent in complex modern urban areas. The problem is especially acute in the United States, where it is estimated that at least two-thirds of the population lives in a fifty-mile-wide coastal belt—and it is not difficult to predict that this percentage will increase markedly in the decades to come. It is not surprising, therefore, to find that oil companies seeking to develop offshore areas for commercial oil and gas production will become involved in problems with a wide variety of state and federal regulatory agencies, particularly those charged with the responsibility of protecting commercial and sport fisheries, marine transportation services, and pollution control.

It should be pointed out, however, that steps taken to solve these problems posed by such regulatory bodies sometimes actually result in lower rather than higher exploratory costs. One specific example is a rule promulgated by a state fisheries agency which does not permit the use of dynamite in geophysical seismic operations off the California coast. Subsequent research conducted by oil and service companies to obtain other usable seismic energy sources has resulted in the most effective use of nonexplosive energy sources. These sources are actually much cheaper than conventional dynamite, and in many ways are also safer in marine operations. In addition, unexploded dynamite may be washed up on the beach, possibly resulting in personal injury. For this reason, states in which dynamite may still be used in offshore explorations require that marks be placed on the dynamite to identify its specific user; it is then possible for penalties to be assessed on those who might inadvertently let loose some unexploded dynamite charges. This regulation also encourages proper handling of explosives at sea to avoid such occurrences.

From the aesthetic standpoint, there are state regulations, particularly on the West Coast, concerning the outward appearance of offshore drilling rigs and the platforms upon which they are built. Stringent regulations are also in effect regarding the lighting of offshore drilling platforms and barges to avoid collisions with surface craft, as well as the use of foghorns, when they are necessary, for navigational safety at sea.

COMPETITIVE ENERGY SOURCES

No discussion of the ocean's energy, particularly when emphasizing economic aspects, would be complete without at least a brief mention of competitive sources, both on land and at sea. In addition to the petroleum reserves on land, there are other sources, such as tar sands and oil shales, that have increasing future potential. These occur in vast quantities in both Canada and the western United States. Experiments have been under way for some time to develop methods to permit their use as a source of oil at competitive prices, and progress is being made, especially with Canadian deposits. Eventually they will furnish at least an appreciable portion of crude-oil production. The major problem is not so much the technology as the high capital investment required per barrel of oil produced.

Coal was the first major source of energy in the country and continues to supply a significant portion of our total energy demands. Recent advances in transmitting electrical power at extremely high voltages over long distances has enabled coal to become more competitive. When the mines are located near large industrial centers, the coal is burned at the mine to produce

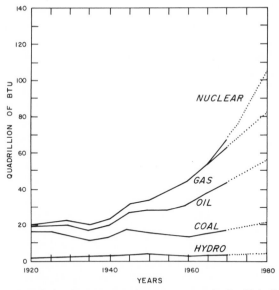

Figure 38. Relative importance of energy sources in the United States

steam to generate electricity, which is then transmitted to major centers of consumption by advanced electrical engineering techniques. This is being done more and more in the eastern and central parts of the United States and has resulted in an increase in the use of coal, particularly in mines where conventional methods of coal distribution would not be economically competitive with oil and gas.

Water power is perhaps one of the oldest continuous power sources in this country. However, it has become less important, percentagewise, as one component of the supply of total energy requirements, although its actual contribution has remained essentially steady throughout many, many decades. Strictly speaking, energy from coal from the continental shelf has been available for some time from mines where coal seams were first mined on land but are continued beneath the sea.

The relative importance of these basic energy sources in this country over the past fifty years and projections for the next decade, with nuclear energy included, are shown in Figure 38, which was prepared from recent estimates of the United States Department of Interior. The most rapid rise of the various components of energy sources is nuclear energy, with water power holding level and the other major sources increasing but less rapidly than nuclear power. This rapid projected rise in nuclear power holds many implications in the area of multi-user problems.

Turning again to the sea, and looking toward the future, for potential energy sources, tides, surface waves, internal waves, and even microbiological sources may make varying degrees of contribution to man's energy requirements. Energy from tidal motion has long been studied and considered as a possible energy source. In the early 1930s various engineering and economic studies were made to determine if the tides in Passamaquoddy Bay in Maine could be harnessed economically; the project was never completed, perhaps because of the distribution cost necessary to find an adjacent market capable of absorbing this large amount of power. Had the advances in electrical engineering that now permit the transmission of electric power at low cost been available then, or had there been large centers of consumption near this tidal basin, the project might have come to fruition.

The primary prerequisites for the successful use of tidal energy include a maximum range in tide through a geomorphic feature requiring a minimum of damming to harness this power. There are many places along the world coastlines where these physical criteria are met. But, again, the requirement of the proximity of major areas of consumption, as well as the large capital investment necessary to produce a kilowatt hour of power, precludes the widespread use of this energy source in the foreseeable future.

However, at the moment there is one power-generating facility utilizing tidal energy. It is situated in the Rance Estuary of the Britanny Coast on the English Channel, where there is a difference between the level of consecutive high and low water of as much as 13.5 meters. During the equinoctial springtide, the maximum flow in the estuary during flow or ebb reaches 18,000 square meters per second. This is about three times the flow of the Rhone River in flood. The point where the tidal power station is located in the Rance Estuary is about 750 meters wide. Production was started in January 1961, and the power plant was placed in operation in 1966. It provides a total power output of 608,000 kilowatts, of which 64,000 are used for pumping.

The harnessing of power from surface waves has long been considered, but no efficient method has been achieved. One of the principle problems is providing sufficient storage of the energy, because of the relatively small and/or infrequent increments in which the power would be generated by the wave action.

More recently, some thought has been given to harnessing the power inherent in internal waves of the ocean. It has been discovered that these waves in many areas sometimes have tremendous amplitudes, at times well over one hundred feet, and periods varying over a broad spectrum from a very short period, of the order of minutes or less, to those related to tidal periods. Here again, the interval between the concept of an idea to harness the energy in these waves and the ultimate practical realization requires a great deal of time, thought, money, and capital investment.

The idea of obtaining energy in significant amounts from microbial action is in the same realm as from waves. However, it is conceivable that under specialized environmental conditions and uses these latter two potential energy sources may some day be utilized to solve many specialized practical problems.

In conclusion, then, it is evident from this discussion that the ocean is important and will become increasingly more important as a source of energy to satisfy the needs of society. The origin of this energy can take many forms, but at the moment the major significant source that can be made readily available in economic competition with other types is oil and gas produced from the continental shelves. In this regard, it should be mentioned that economists believe the projected demand over the next twenty years for petroleum will require the discovery of an additional 89 billion barrels.

Much remains to be done before the other potential sources discussed can begin to make any significant contribution to our total energy needs. However, under certain specialized environmental conditions and needs, some of these sources will eventually be important, at least qualitatively, if not quantitatively, in the total energy demands.

Future exploitation of oil resources from the continental shelf in economic competition with the other sources described must be achieved for the reasons discussed. It is imperative, therefore, that sufficient emphasis be placed on continuing research and development in basic oceanography and ocean engineering if the myriad technical and scientific problems are to be solved, as we continue to look to the sea as a major source of energy to satisfy the ever-increasing needs of our world society.

Resources from the Sea

JAMES A. CRUTCHFIELD

SOME unknown realist once made the comment that "men have always been fascinated by the sea—particularly those who do not have to make a living from it." Much the same tone pervades the analysis of marine resources and their present and prospective contribution to human welfare. Intermingled with estimates of the very real present benefits derived by man from the living and nonliving resources of the sea are clouds of mythology, irrelevant (but very large) numbers, and a misleading tendency to confuse need with demand and physical availability with cost-competitive supply. General public discussion has tended to ignore the tremendous differences among marine resources in terms of current economic contribution, of conditions under which they are obtained from the marine environment, and of the technology required to convert them to useful goods and services. There is no unified "bundle" of resources to be had from the sea. Rather, there are groups of presently or potentially useful raw materials, some of them extensions of land-based resources, some living in or dissolved in sea water, and some lying on or under the seabed at great depths. Some are already of major importance; others are barely at the margin of economic production; many await the development of yet unknown methods and materials before they will be of economic interest to man.

The incredible flexibility with which a modern industrial economy uses raw materials inputs leads to little-understood implications. Improved technology permits more extensive and intensive use of given materials; substitution possibilities grow like the amoeba; and the capacity to transform inputs, chemically and physically, into desired forms has proceeded equally rapidly. This increase in man's ability to rearrange the earth's surface greatly reduces the impact of any single resource discovery. In technical terms, the net benefit from a new resource option is always bounded by the

James A. Crutchfield is professor of economics at the University of Washington.

cost of the next best alternative source, and the numbers and desirability of alternatives are constantly growing. Thus, developments that make it economical to use living and nonliving resources of the sea will usually appear as changes in relative costs of production rather than in dramatic alterations in the entire economic structure of a region or country. Finally, the difficulties of the domestic fishing industry and the latent character of most of the mineral resources of the sea have lent renewed support to a disturbing wave of protectionism, a demand for national self-sufficiency, that must be regarded as a major threat to the industrial vitality of the American economy.

The following discussion of marine resources therefore starts from the following basic premises: (1) long-term subsidies and trade restrictions are not acceptable means of development or rehabilitation of resource-oriented marine industries; (2) the United States must not be confronted with critical shortages of raw materials on short notice; and (3) resources in the marine environment should normally be developed by private enterprise, with assistance from government programs and policies directed toward increasing economic efficiency and encouraging technical progress and innovation.*

THE APPROACH

The principal resource components of the marine environment consist of "fisheries" (used here as shorthand for all living resources), minerals dissolved in seawater, minerals on and beneath the sea floor, oil and gas, direct power, and fresh water. In addition, inshore and estuarine areas generate a flow of recreational services that now provide the base for a number of important industries.

For each of these resources, I shall present a summary of demand projections; of prospective prices and supplies, over the long run, from land-based sources; and of the present state of the industry exploiting the resource, with emphasis on technology and organizational structure. Within this framework it is possible to distinguish those marine resources that are presently competitive and expandable, those expected to become profitable within a short time, and those that will become competitive only through major changes in demand, in exploration and exploitation capability, or in both. Brief reference is made to the critical importance of the legal environment in which both living and nonliving resources are developed and used by man. In particular, the absence of property rights in the conventional sense of the word poses a major obstacle to both national and international development. As suggested elsewhere in this volume, the most important institutional changes

* Much of this material is drawn from the report of the Panel on Marine Resources of the Commission on Marine Science, Engineering and Resources, prepared by the author (fisheries section), Dr. Creighton Burke (oil and gas section), and Dr. John Albers (fresh water, power, and minerals). This does not imply endorsement of the commission as a whole or by individual members.

that must occur before the full potential of marine resources can be realized involve the legal framework defining access to them and the protection of the right to exploit them.

THE MINOR COMPONENTS

Power

No one, layman or scientist, who has observed the sea in motion can fail to be impressed by the power it generates. One of man's oldest dreams has been to harness even a fraction of that enormous energy to relieve pressure on fossil fuels and on increasingly scarce locations for low-cost hydroelectric power. Unfortunately, reality is less exciting than the dream. Of all the energy expended by the sea, only tidal movements appear to offer prospects for power generation in the near future, and experience to date with actual projects and detailed studies of other potential sites indicates that costs are much too high to make them feasible. Moreover, the emerging technology of nuclear power generation, particularly where it is coupled with other useful outputs (including desalination of brackish or salt water), clamps a price ceiling on any power source from marine energy. For the moment, it would be difficult indeed to justify a major research or development program devoted to power from the sea.

Fresh Water

The conversion of desert into garden by the desalting of seawater is an equally old dream, but one which shows much greater promise for the future. Desalination is, of course, an accomplished fact. Several technically sound methods have been developed, and plants are now operating in various parts of the world. The question is not whether fresh water can be produced from the sea, but whether it can be produced at a cost competitive with that from other sources (including interbasin transfers, the recycling of waste water, and possibly the modification of weather to increase precipitation or to improve its timing).

The physical supply of water is completely unlimited, since the hydrologic cycle is barely affected by man's activities. The problem is to provide water of appropriate quality at lowest cost in the quantities and at the places where man wishes to make use of it. In these respects, desalination appears to offer much promise for the future. The principal alternative, particularly in the more arid parts of the United States, requires massive interbasin transfers of water, involving expenditure of billions of dollars on long-lived capital equipment. This approach locks whole regions of the nation into a rigid system of fresh-water outputs over long periods of time. Since any forecast of the rate and pattern of economic and demographic change is imperfect even over a range as short as five years, there is real danger that massive investments will go into delivering the wrong quantities of water to the wrong places. Desalination offers much greater flexibility, at least with respect to the coastal areas, and opens up the attractive option of supplying fresh water in small increments. For many locations, the marginal flexibility of supply made possible through desalination may provide an important supplement to systems making use of surface and ground water.

Despite the unquestioned importance (and reasonable magnitude) of efforts to derive fresh water from the sea in quantity and at competitive prices, the goal remains just beyond reach. Even today, it is impossible to obtain a firm estimate of production costs without specifying a technology certain to be outmoded before a new operating plant could be brought into production. Moreover, estimates are blurred by the unavoidable necessity of allocating common costs of a nuclear plant between water production on the one hand and power on the other—a combination which now appears to be the most economical technique for desalination. It is safe to say that present technology, coupled with developments already foreseen with some confidence, establish a ceiling price of no higher than seventy-five cents per thousand gallons. Since this is far from competitive with most surface supplies, it would not make irrigated agriculture profitable in most locations. However, it is well within the realm of practical application for many isolated arid areas, and would be quite attractive as a source of incremental supplies for small- to medium-sized cities.

From a longer-range point of view, cost estimates of twenty-two cents to thirty-seven cents per thousand gallons are cited by presumably competent research engineering firms, and even the midpoint in this range would make desalted water very attractive by comparison, for example, with the probable cost of Columbia River water delivered to the Southwest. In addition, fresh water obtained from the sea would largely avoid the problem of uncompensated economic costs imposed on the area of origin by any interbasin transfer project. In general, further cost reduction in the production of desalted waters is related to two factors: the deposit of mineral substances incurred in the flash distillation process, and—possibly but not certainly—the ecological impact of the discharge of heated and somewhat concentrated brine. The first problem is a subject of scientific and engineering inquiry which is going on apace in a well-defined American program shared by federal, state, and university personnel. The second obstacle presents more difficult problems of concept and quantification, but a systematic analysis in economic and social terms of alternative locations should make it possible to identify safe discharge locations and to assess their relative cost advantages.

It might be noted, parenthetically, that the concentration of brine in the process of removing fresh water from sea water may enable the further economic recovery of dissolved minerals. Whether the cost effect of concentration at levels now contemplated in desalination would be sufficient to offset other location factors is problematical, however.

In summary, one might mark this dream as sufficiently close to reality to warrant continued (and probably expanded) research efforts. It should not be regarded, however, as a crash program to prevent premature death by thirst of hundreds of thousands of southern Californians. It should be placed properly among the family of alternative techniques for providing additional supplies of water of proper quality at locations where its net value will be maximized.

THE HISTORICAL MAJOR COMPONENT: LIVING RESOURCES

Although the public fancy has been caught by the spectacular growth of offshore gas and oil production and the titillating recital of huge numbers purported to describe the hard-mineral potential of the sea, the fisheries, broadly defined, still provide much the greatest total value output from the sea on a world-wide basis. That petroleum has now outstripped the fisheries in the United States is due not only to American technical dominance in offshore petroleum but to the stagnant state of the American fishing effort.

Unlike the handful of mineral producers now operating in the marine environment, fishing is an old and mature industry, with a history extending back to the beginning of recorded time and an extensive scientific literature. As in the case of marine resources as a whole, generalizations about "the fisheries" can be dangerously misleading. Despite justified concern over the poor performance of the American-flag fisheries and mounting evidence of international conflict and excessive pressure on some marine populations, the fact remains that the output of marine fishery products has been expanding by more than 6 percent per year since the end of World War II. Moreover, the disappointing record of the American effort in recent years should not blind Americans to the fact that revolutionary changes in detection, harvesting, processing, and marketing of fish have taken place in the last twenty or thirty years.

These spectacular growth figures must be treated with caution, however, since decomposition of the gross magnitudes reveals some details that are worrisome. For example, almost ten million tons of the total postwar increase is accounted for by the spectacular growth of a single fishery, the Peruvian anchovy operation; and other major portions were provided by the equally sharp expansion of the South African fish-meal industry and the emergence of the Soviet Union as a fishing power in the North Atlantic and North Pacific. The wide-ranging tuna fishery, expanding continuously in response to a steady growth in consumer demand and a technologically progressive fishing operation in the major countries concerned, is now beginning to show signs of economic trouble ahead. A recent FAO task force report suggests strongly that the world tuna fisheries may be approaching a plateau beyond which further effort will not produce increases in output. The great demersal fisheries of the North Atlantic have already reached a state of full, or nearly full, utilization of the commercially desirable species, and the productive trawling banks off the west coast of Africa are rapidly approaching the same state.

In short, the great expansion of the postwar years has been achieved largely in continental-shelf waters (with the exception of the tuna operation), using improved versions of familiar gear and operating on species already acceptable in the market. Continued expansion of the sea's contribution to marine food supplies can be achieved, but it will require an accelerated research and development effort to harvest species presently too inaccessible or too scattered to be used profitably, to develop new and desirable products from presently unmarketable species, and to improve the economic efficiency of the individual catching-transportation-processing sequence.

How important are the living resources of the sea to the United States and to the world? According to recent estimates, world landings of fish products are now in the vicinity of 60 million metric tons annually, nearly three times the landings recorded as recently as 1950 (Table 2). Although

TABLE 2
TOTAL WORLD FISHERIES CATCH, 1950–67
(Millions of Metric Tons)

Year	Catch	Year	Catch
1950	21.1	1960	40.0
1951	23.5	1961	43.4
1952	25.1	1962	46.9
1953	25.9	1963	48.2
1954	27.6	1964	52.5
1955	28.9	1965	53.3
1956	30.4	1966	56.8
1957	31.5	1967	61.1
1958	33.2	1968	64.3
1959	36.7	1969	62.9
		1970	69.3

SOURCE: Food and Agriculture Organization of the United Nations, *Yearbook of Fishery Statistics, 1970* (Rome, 1972).

these figures are small relative to total world food production, fishery products make up more than 10 percent of total animal intake and about 3 percent of total protein from all sources. In view of the population pressures that bear heavily on most parts of the world, particularly in the underdeveloped countries, this is far from a trivial contribution to the never-ending fight against hunger.

Moreover, these gross figures conceal the economic and social significance of marine fisheries by region and nation. For some countries, food from the sea represents an indispensable source of animal protein, and any significant decline in fish landings would disastrously worsen the already deficient diets. Virtually all of the African and Asian coastal countries fall in this category. For other peoples, such as the Scandinavians, high per capita consumption of fish is not a matter of life or even health, but of strong tastes and preferences. For many nations, food from the sea is a vital source of foreign exchange; both fish and fish meal are important items of international commerce. In recent years, the development of efficient methods of producing fish meal and its use in a wide variety of animal feed products has spectacularly increased the generation of food from the sea via the indirect process of conversion by land animals.

In brief, even at today's levels of output, the contribution to man's well-being by the living resources of the sea is a matter of deep national and international concern. Efforts to improve the performance of the industries exploiting them carry welfare implications for the nation and the world too important to be neglected.

Demand and Supply

There is every reason to argue that the demand for protein food from the sea will continue its persistent growth and will spread across a broad range of fishery products. Specialty-processed fish and the more exotic types of fish and shellfish are in great demand in the markets of the developed countries. At the other end of the spectrum, demand for low-priced fish continues to respond to increases in both population and income in the less developed nations. The tendency in the higher-income countries for per capita consumption of fish in direct form to level off has been offset by an accelerating demand for animal and dairy products, for which fish meal is an increasingly desirable input. The record of recent years makes it clear that the steady expansion in the demand for animal proteins, whether for direct consumption or through conversion by other animals, will accelerate the search for ways to use previously unmarketable fish and to improve the consumption characteristics of end products.

Supply prospects for marine fishery products are much more difficult to forecast. Despite the increasing interest of scientists and oceanographers in providing systematic appraisals of the ocean's fishery potential, current estimates range from roughly 55 million metric tons annually to 2 billion metric tons—a range hardly comforting to scientists and nutritionists concerned with the need to expand protein food supplies (Table 3).

TABLE 3
ESTIMATES OF TOTAL OCEAN YIELDS OF AQUATIC ANIMALS
(Millions of Metric Tons)

Forecast	Year	Method	Author
21.6	1949	ext.[a]	Thompson
55.4	1955	ext.	FAO
50 to 60	1960	ext.	Finn
55 (bony fishes)	1962	ext.	Graham and Edwards
55 (by 1970)	1962	ext.	Meseck
60 (bony fishes)	1962	ext. ef.[b]	Graham and Edwards
66 (by 1970)	1965	ext.	Schaefer
70 (by 1980)	1962	ext.	Meseck
80	1965	ext.	Alverson
70 to 80	1965	ext. ef.	Bogdanov
115 (bony fishes)	1962	ef.	Graham and Edwards
160	1965	ext.	Schaefer
200	1965	ef.	Schaefer
200	1965	ef.	Pike and Spilhaus
1,000	1966	ef.	Chapman
180 to 1,400	1962	ef.	Pike and Spilhaus
2,000	1965	ef.	Chapman

SOURCE: Schaefer and Alverson 1968, p. 82.
[a] ext. = extrapolated from catch trends or existing knowledge of world fish resources
[b] ef. = energy flow through food chain

One reason for the apparent discrepancy is that some investigators are concerned with total biological production, while others, implicitly or explicitly, are talking in terms of yield or the total product of the biological activity minus those quantities taken by nonhuman predators or used up in the biological process itself. More important, any estimate couched in terms of physical yield ignores the dual facts that man's technical ability to harvest and the demand patterns resulting from tastes, preferences, incomes, and the availability of substitutes will always impose a lower economic limit on yield than the maximum possible physical output.

A more meaningful way to define the functional relation in question would run in terms of a supply function: that is, a plot of the total quantities of fish and shellfish that would be forthcoming at successively higher relative prices. This would provide an estimate of production capability from the sea on the critically important assumption that such production must yield an economic return equal to or greater than that earned from the production of other goods and services with the same capital and labor inputs. In this sense, a supply function geared to present fishing technology, present fishing areas, and present patterns of demand would probably show rapidly increasing marginal costs at an output of perhaps 75 to 85 million metric tons. But this is a purely static statement of yield possibilities in economic terms. On the more realistic assumption of continued improvement in vessels and gear, in preservation, storage, and transportation techniques, and in marketability of end products, the supply function would shift rightward and its slope would be markedly reduced. Viewed in this light, an estimate of 400 to 500 million metric tons per year of economically useful output from the sea does not appear unreasonable.

Expansion of output of the world's fisheries is likely to be accompanied by a steady reduction in the average unit price of the species taken. We are already exploiting at a high rate the more accessible and more desirable species, and order-of-magnitude increases in production can be achieved only by utilizing fish from less accessible grounds or by taking species which produce end products of lower value. This is, of course, nothing new in the natural resources field. The remarkable ability of the American economy to maintain or even to reduce the relative cost of raw material inputs to its rapidly growing gross national product is almost entirely attributable to the pace of technological development which has permitted substitution of poorer quality inputs as the more valuable resources have been used up.

In summary, the world is already making relatively good and important use of the living resources available in the sea, but it is far from achieving the full potential that will unfold in the normal course of economic, engineering, and scientific progress. The quantities and values involved are unquestionably so large that the careless manner in which these resources have been exploited and managed during the past century can no longer be tolerated.

The All-American Problem

From the standpoint of the American-flag fishery alone, the prospects are much less sanguine. Demand for fish products for direct consumption in the United States has remained remarkably stable, in per capita terms, over the past forty years. Nevertheless, the growth in population, coupled with the remarkable increase in the use of fish meal for animal and poultry feed, has produced a rapid increase in aggregate consumption of all fishery products. The United States is now the largest single consumer of marine seafood products, using approximately 12 percent of total world landings.

Despite this stimulus to the American fisheries from the standpoint of demand, the production response of American vessels has been most disappointing, as indicated in Table 4. Total landings have remained virtually static over the past thirty years; the entire increase in total consumption from 1950 to 1971 was accounted for by increased imports. In addition,

TABLE 4
UNITED STATES CATCH OF FISH AND SHELLFISH, 1945–67

Year	Landings (millions of pounds)	Value (millions of dollars)	Year	Landings (millions of pounds)	Value (millions of dollars)
1945	4,598	270	1958	4,747	373
1946	4,467	313	1959	5,122	346
1947	4,349	312	1960	4,942	354
1948	4,513	371	1961	5,187	362
1949	4,804	343	1962	5,354	396
1950	4,901	347	1963	4,847	377
1951	4,433	365	1964	4,541	389
1952	4,432	364	1965	4,777	446
1953	4,487	356	1966	4,366	472
1954	4,762	359	1967	4,054	440
1955	4,809	339	1968	4,160	497
1956	5,268	372	1969	4,337	527
1957	4,789	354	1970	4,907	613
			1971	4,969	643

SOURCE: Fish and Wildlife Service, Bureau of Commercial Fisheries, U.S. Department of the Interior, *Fishery Statistics of the United States, 1969, 1971.*

while landings of some high-valued species have expanded, most of the growth required to compensate for reductions in catch from overexploited fisheries has come from relatively low-priced species. Thus, the increase in the value of the American catch over the past twenty years has barely kept pace with the general increase in the American price level. Direct employment in the United States marine fisheries was only about 132,000 in 1969, and a substantial proportion of these fishermen are only part-time participants.

This unimpressive record is even more startling in view of recent assessments of potential yield from continental shelf waters of the United States. These studies suggest that the total potential yield in physical terms is nearly ten times current American production. Moreover, foreign-flag vessels are now taking approximately as much fish from American continental shelf waters as our own fishermen are landing.

The reasons for this dismal performance are complex. Some of the problems are attributable to the nature of the resource itself. Marine fish populations are found in a difficult environment, frequently widely dispersed, under conditions that make direct observation impossible. In addition, many of them are biologically susceptible, in greater or lesser degree, to depletion or even to destruction by overexploitation. Man has taken only a few tentative steps toward a harvesting concept; for practical purposes, world as well as national fisheries remain a hunting operation.

The major impediments to continued development of the American-flag fisheries are deeply rooted in the structure of this government. Despite considered legal opinion that the federal government has authority to manage the fisheries within the three-mile limit, as well as in the "shadow zone" from three to twelve miles, it has not chosen to do so except where the United States is party to an international fisheries agreement. By long tradition, jurisdiction over marine fisheries has remained largely the province of the individual states, with the federal function severely restricted to service operations. The essential tasks of management, research and development, statistical data collection, and protection against environmental damage must be performed by or through state agencies.

The results have been disastrous. In terms of both economic and political theory, it is essential that resource management be conducted in an organizational framework where the decision-making authority is coextensive with the range of the alternatives to be considered. Clearly this prescription is violated in most of the important fisheries of the United States, as well as in the international arena. Even where states have been able to join in regional compacts, the absence of a common set of regional objectives has severely limited their effectiveness. It is hardly surprising, then, that the states have generated an ever-increasing mass of restrictive legislation, most of it clothed in the shining garments of conservation, but bearing the clear marks of pressure politics. An overwhelming proportion of regulations affecting fisheries in the individual states, whether statutory or administrative, reflects power plays by one ethnic group of fishermen against another, by owners of one type of gear against another, or by fishermen of one state against those of another state.

The quantity and quality of the research and development work varies widely from state to state, and the important statistical series relating to catch, effort, and other significant records of fishing activity are dissimilar in coverage, incomplete, and of dubious accuracy in most cases. Although the National Marine Fisheries Service has continuously tried to coordinate and plug gaps in state programs, the aggregate national fishery effort in the public sector remains fragmented and basically unsatisfactory in level and composition.

The situation is further aggravated by the structure of the federal fishery function, the background of its leadership, and the geographic dispersion of its activities. It is not surprising that most of its top personnel are drawn from the field of fisheries science; but this composition has resulted in a strong orientation toward biological research, with much less emphasis placed on the technological and economic aspects of the industries that use the resources. The competence of the National Marine Fisheries Service in biological and some technological areas is not matched by capability in vessel and gear technology, processing and marketing, or economic analysis of local, national, and international factors affecting the fishing industry.

Thus, after decades of operation, the service is still unable to define the magnitude and the temporal and area distribution of fish stocks with sufficient precision for minimal investment planning by industry. The service has achieved some remarkable break-throughs in gear research; but the number and significance of these accomplishments is pitifully small because of totally inadequate budgeting and the fragmentation of this work among many field offices. Finally, the service suffers continually from its susceptibility to external pressures. Because the fisheries are found around the entire vast perimeter of the American shoreline, pressure on the service to "do something for the local fisheries" results in a dispersion of funds and personnel that makes it very difficult even to plan, much less to implement, bold and imaginative new programs, or to concentrate effort in areas of greatest promise.

Lacking a clear-cut set of objectives and a mandate to pursue them, the federal fishery function has been built up, element by element, from disparate projects and programs of limited scope, most of them in response to specific pressure from legislators or industries. It is estimated that no less than 80 percent of the bureau's current $50-million-a-year budget is locked up in projects which may or may not have continuing usefulness, but which cannot be discontinued in favor of challenging new opportunities.

Who Owns the Sea?

In both the national and international spheres, man's ability to utilize optimally the wealth available from the sea has been severely handicapped by the peculiarities of property rights in the marine environment (for a discussion of this common property problem, see Christy and Scott 1965). Regardless of the form of economic organization—market-oriented or socialist —no fishing industry can make efficient use of a marine resource unless it has the ability to make appropriate decisions as to intertemporal harvesting rates. Where a fishery is subject to management by a single owner, it will obviously tend toward an optimal position in which the maximum flow of money benefits net of associated operating costs is achieved. Given the biological characteristics of fishery resources, this requires careful consideration of the impact of today's harvesting activities on the flow of raw materials available in later periods. In the absence of sole ownership, there is obviously no incentive for any single fisherman or group of fishermen to curtail current activities to achieve an optimal time distribution of catches, because no property right guarantees that the fish not taken today will be available in large quantity or at greater weight in the future. What the fisherman does not catch today simply goes to other fishermen.

Fisheries throughout the world have therefore been characterized by a persistent (and in some cases disastrous) tendency toward overexploitation. At a minimum, this results in excessive amounts of capital and labor tied up in essentially duplicative harvesting activities. In worse instances, this has been accompanied by severe depletion or even destruction of the populations concerned. The process is not self-correcting; and if the value of the catch relative to the cost of harvesting is sufficiently high, the virtually certain result is severe physical depletion as well as unnecessary economic cost. And the situation is constantly aggravated as the range and sweep efficiency of modern fishing vessels increases and as some nations have developed the capacity to integrate analysis of oceanographic parameters with the operations of complex fishing fleets.

We simply do not have the time now to accumulate, slowly and painfully, the data required to estimate yield potentials from a fish population and then set about, in equally leisurely fashion, to erect a management program to prevent excessive rates of exploitation. Today's fishing techniques are more likely to produce explosive growth in fishing effort whenever new opportunities appear. Nor have political or administrative techniques yet been devised on the international or national level to prevent the inevitable "overshoot" which leaves the industry with heavy overcapitalization, continuing losses, and a short-run incentive to continue excessive rates of fishing.

It should be stressed that this somber analysis is not inconsistent with the observed fact of rapidly rising fish landings from marine resources during the past twenty years. It simply implies that the ability to expand effort, extensively and intensively, must be finite and that we are increasing landings only by swamping the losses from overexploited populations with catches from newly discovered or newly developed sources.

The typically small scale of fishing enterprises suggests an important role for government in the exploration and definition of exploitable populations, gear research and development, and the basic testing of new concepts of harvesting. All of these activities promise to extend significantly the quantity, quality, and reliability of marine food-resource supplies; but none of them is likely to be done (or to be done to the proper scale and in the proper manner) by individual units that could capture only a small fraction of the total benefits generated.

Nor are these activities likely to achieve their potential benefit unless the resulting improvements are accompanied by an equally solid development of man's ability to control entry into the fisheries and thus to approximate the optimal rate and composition of fishing effort. This will require drastic revision of widely held concepts of the appropriate objective fishery management, both nationally and internationally; and since no single system of management will be optimal for all individual members of internationally shared fisheries, the problem of reaching a satisfactory common ground becomes even more difficult. Nevertheless, almost any compact must offer improvement over the results achieved in the past (and to be expected in the future) from a destructive unrestricted race to catch fish before others do. Even a distinctly second- or third-best international agreement to rationalize operations (that is, to use only that amount of gear required to take the desired catch) would yield tremendous returns. As in so many areas of natural resource management, man's technical and scientific competence appears to have outdistanced by far his ability to assure orderly development and usage.

THE NEW MAJOR COMPONENTS: OIL AND GAS

The future prospects for crude petroleum and gas * from the sea look as bright as the immediate past record of performance. Offshore petroleum production is now so familiar that it is worth recalling that the first real physical exploration beyond sight of land was conducted in the Gulf of Mexico in 1944, and the first well drilled beyond sight of land came into production in 1946. It was not until 1951 that the first offshore pipeline was completed. Since that beginning, the petroleum industry has moved more and more rapidly to resolve both exploration and production problems. Over the last ten years production capability has moved from less than seventy feet under water to more than three hundred feet. Pipelines have been laid successfully at that depth, and it is generally accepted that production wells can now be drilled in water six hundred feet deep or more (see Taylor 1969).

Space precludes a detailed analysis of the demand for oil and gas; nevertheless, a few basic figures will serve to explain the tremendous drive for discovery and development of offshore oil throughout the world. We are in the midst of an unprecedented surge in the demand for energy. It has been estimated that during the next twenty years oil consumption in the Free World nations alone will rise by more than 250 percent. Estimates by the United States Bureau of Mines predict an increase in world demands for petroleum of 120 percent by 1980 and 355 percent by the year 2000. Thus, even though petroleum is expected to provide a smaller share of future energy requirements in the United States, the consumption of petroleum will continue to rise rapidly in absolute terms.

It is simply impossible to summarize in a few words the overall supply prospects for petroleum in order to evaluate properly the role of offshore

* Although natural gas is an important marine resource, it is normally found as a by-product of the search for crude oil. In the interest of brevity, the following discussion is confined to petroleum from the seabed.

production. Suffice it to say that neither the world nor the United States is facing any serious threat of petroleum shortage, as would be evidenced by a rising marginal cost of supplying crude petroleum inputs to a growing national and international output. This might seem surprising, in view of the enormous drains on proved reserves in the years since World War II. It can readily be explained, however, in these terms. Proved reserves that can be developed profitably at present prices with presently known technologies are substantial, though reserves available in the United States, defined in this way, have been declining as a proportion of current output. Another block of crude petroleum would be available at higher prices (or at current prices if technological progress could be accelerated even slightly). A third and much larger block would be available if new technologies could be developed and other problems, many of them involving protection of environmental quality, could be resolved. This last block includes oil from shale and tar sands.

Even with these relatively large land-based supplies, however, discovery of new oil fields in offshore waters is tremendously attractive from a profit standpoint; and, given the highly concentrated structure of the oil-refining industry, competition for reserves that guarantee market shares is a compelling force driving major firms in the United States and elsewhere into the sea.

Any estimate of reserves on the continental margins is obviously speculative, and the potential in deeper waters is even less amenable to sensible estimating procedures. Even so, it is apparent that commercially usable supplies, at present technological levels, are available in quantity on the continental margins throughout much of the world. Recent exciting discoveries in the Caribbean raise the hope that even the deep-sea depths may contain important supplies that will eventually yield to the march of technological progress. Even today, production of offshore oil is under way in the waters of twenty-two countries, and more than seventy-five nations on five continents are undertaking, or have granted permission to undertake, exploration off their shores. Some 16 percent of the world production of crude oil is now coming from offshore sources, and about 20 percent of the proved petroleum reserves are in these areas. It is anticipated that by 1980 at least one-third of total world production will come from the continental shelves and slopes.

The magnitude of this activity in terms of private investment is equally impressive. Led by American firms, private corporations are investing at a rate of more than $1 billion a year in offshore petroleum activities, and these expenditures have been growing at a rate of nearly 18 percent per year. More than $7 billion has been invested by the petroleum industry in the waters off the United States alone.

Four separate sets of capabilities are required for economic extraction of petroleum from the seabed: exploration, drilling, production, and transportation. The rate of progress in each of these areas has been most impressive. Capability has developed most rapidly in the field of seismic exploration, which in most respects can actually be carried on more effectively at sea than onshore. A whole new technology has permitted the industry's offshore drilling operations to move from a depth of seventy feet ten years ago to more than six hundred at the present time. There is little doubt that exploratory and production wells can now be drilled successfully in water as deep as one thousand feet.

Production capability has proceeded less rapidly. A completed well requires considerable periodic maintenance under any conditions, and these needs are much more drastic at sea. In addition, the greater protection required against offshore blowout and leakage is emphasized by the recent disaster in the Santa Barbara channel. To date, physical requirements for production from offshore wells have been met by installing the basic equipment on permanent platforms resting on the sea floor and extending above sea level. This limits the total operating capability to relatively shallow waters —roughly three hundred fifty feet or less at present. Eventually it may be possible to complete wells entirely underwater on a regular basis, but such broad-based technology is not yet at hand.

Finally, offshore production of petroleum requires solution of both technical and legal problems if crude oil is to be transported from well to refinery economically and with a minimum of interference to other activities in adjacent areas. In general, cost factors present the principal obstacles, but even now oil can be transported economically from any depth where production is currently feasible.

The Problem of Offshore Oil

The discussion thus far would suggest no problems with respect to the development of a highly important increment to world and national oil supplies from offshore sources. The scale economies involved in onshore oil production have given rise to an industry dominated by large, technically advanced, research-oriented firms. These have brought to the depths of the sea both an attitude and a set of skills admirably suited to rapid solution of problems involving materials, personnel support systems, and the large amounts of capital required to explore and produce under conditions of great risk and uncertainty.

There are, however, problems indeed; and, as in the case of the fisheries, they are more economic and political than scientific or technical. The petroleum industry suffers from its own version of the common-property dilemma discussed in connection with marine fisheries. The result has been a persistent tendency toward overproduction. In the United States, the formulation of a so-called conservation program has adversely affected the efficiency with which capital and labor are employed in the industry, and has led to production and import restrictions that raise serious questions as to the long-term national interest.

This is a topic for another book rather than another paper, and it obviously cannot be dealt with adequately in this essay (for a definitive treatment, see Lovejoy and Homan 1967). In essence, American policy has proceeded on the assumption that it is unsafe on security grounds to rely on foreign oil for any significant proportion of total domestic consumption, despite the fact that lower production costs would permit foreign oil to be delivered at prices ranging from $1.25 to $1.50 per barrel lower than those now prevailing in the United States. This artificial enhancement of American domestic crude prices is an integral part of the prorationing approach to conservation regulation. It derives additional support from the massive subsidy it provides for hundreds of thousands of stripper-well owners and for large integrated producers who are able to take advantage of the more than generous depletion allowances and other tax concessions afforded crude-oil producers in this country. Price levels in the United States are so far out of line that the domestic restrictions on imports, which now limit foreign sources to approximately 20 percent of the domestic market, are essential to the system.

The political strength of the petroleum industry is such that these policies have never been subjected to rigorous public scrutiny, despite the fact that most economists who have reviewed the situation have argued strenuously for an entirely different set of policies (Lovejoy and Homan 1967; see also McKie and MacDonald 1962 and Davidson 1963). With respect to the national defense argument, for example, it is difficult to see how long-term American security is enhanced by policies that subsidize uneconomically rapid production from supplies under direct political and physical control of this country. If, on the other hand, we are concerned with the impact of nuclear war, it would appear that the future petroleum supply picture would be the least of our worries.

The standard national defense argument for a program of quota protection and heavily subsidized domestic production of crude oil is the "industry in being" concept; that is, reliance on foreign supplies of crude oil would subject us to the danger of military defeat or industrial decay if access to the politically unstable economies of the Middle East and of Latin America were blocked. The oil lobby is notably silent on the parallel argument that crude petroleum is only the first step in the production process through which the natural resource is converted into industrial inputs; unless duplicative transportation and refinery facilities were assured (presumably "hard" enough to withstand nuclear attack and sabotage), the additional crude capacity would be of no use whatsoever. Meanwhile, it would appear that Americans are paying a price running into billions of dollars a year for the dubious privilege of using up our own oil faster than market conditions alone would dictate.

The implications of this situation for offshore oil development are obvious. To keep domestic production from flooding American markets, it has been necessary to hold onshore production well below capacity, and this situation will not be altered for at least a decade. Any increase in offshore production must somehow be fitted into the prorationing scheme—a problem of no small political proportions, and one that will become increasingly touchy as the new oil discoveries in Alaska come into production. At present, there is a strong incentive to develop offshore supplies, even in the face of redundant capacity in the American industry, as a defensive measure, to insure against the disaster of inadequate supplies for tremendously expensive refineries in the future. The present leasing policy on federal- and state-owned offshore oil lands also tends to encourage rapid exploitation, since it is based on payment of a bonus coupled with a percentage royalty on subsequent production. Since these bonuses are substantial, amounting to more than $600 million in the Santa Barbara channel leases, the lessees are under heavy pressure to develop production as quickly as possible to recover the initial outlay.

All of this adds up to a gnawing concern that the problem facing the American government is actually not one of stimulating offshore production but rather one of assessing, in long-run terms, the risks of using cheaper oil from foreign sources, at a very considerable saving to the American economy, while reserving part of America's own offshore reserves against future needs. Various schemes might well be developed that would encourage exploration without the necessity of immediate production to recover funds invested in the process. Obviously, the ramifications of these issues go far beyond the matter of proper policy toward offshore oil. It would seem that a major review and assessment of American petroleum policy, as it relates to both national and international sources, is long overdue; and I would venture the opinion that the interests of the American people as a whole would be well served by a major alteration in federal and state policy toward crude-oil production and imports.

HARD MINERALS

As in the case of living resources and petroleum, the hard-mineral resources of the sea are best analyzed in terms of potential demand for minerals from all sources and of the contribution that can be made to meeting that demand from the marine environment. Again, the relevant question is not whether the United States can meet its "needs" for minerals, but whether it can obtain supplies sufficient to meet rising demands without significant increases in relative costs. The nation must also be aware of the possibility that specific minerals or groups of minerals of key importance for defense or industrial purposes might suddenly become critically short if present overseas sources were cut off for political or military reasons.

Two conclusions stand out clearly from demand figures projected to 1985 and 2000. First, the projected increase in demand over the next three decades is very uneven. Second, the aggregate demand for minerals as a whole will increase very substantially throughout the period. In short, a continuation of satisfactory levels of growth in the developed countries of the world implies a rapid rate of increase in the demand for minerals of all types, with selective demands calling for truly heroic efforts to increase the supply of some products. This conclusion is, of course, modified by the amazing ability of the developed economies to substitute abundant for scarce materials and nonmineral for mineral inputs; but it is apparent that the mineral industries of the Free World, including the United States, will be called on to meet major increases in demand.

At first glance, it would seem obvious that mineral resources, exhaustible and nonrenewable, cannot be developed at a pace that will match these demands, and that the exhaustion of the best deposits and the need to move to less accessible and lower quality sources will inevitably contract production possibilities in all parts of the world. When we consider that the United States, during the last thirty years, has used more minerals and fossil fuels than the entire world used in all previous history, the magnitude of the apparent threat becomes even greater. Moreover, from a purely national point of view, forty of seventy-two strategic and critical commodities in the mineral field, indispensable to the American economy in times of conventional warfare, are at present largely imported from foreign countries, and this dependence will increase over time.

This forbidding view of the future, however, is based on the assumption of a static and essentially nonprogressive world. Once the fact of technological progress is accepted, the apparently irrefutable argument leading from increasing resource scarcity to diminishing per capita returns in economic output and in human welfare is not even a half truth. This is not black magic. It simply implies the systematic substitution of the less scarce for scarce products, of labor and capital for resources, and of nonmineral for mineral raw materials. This process is not by happenstance, but is a logical response to the very fact that scarcity threatens. The growing sophistication of international trade and its impact on better world utilization of resources operates in the same direction. The stimulus to research and development becomes endogenous rather than exogenous, thus holding back the ghost of Malthus. These factors account for the apparently anomalous results obtained by Barnett and Morse, which suggest that in the United States the costs of extractive output have actually declined for all natural resources (with the possible exception of forest products) from 1870 to 1957 (Table 5). Even more intriguing, the costs of extractive goods relative to nonextractive goods have declined over this period, and the decline has been greatest in the exhaustible resource area (minerals) and least evident or absent in a reproducible resource (forest products).

TABLE 5
LABOR-CAPITAL COST PER UNIT
OF NATURAL RESOURCE OUTPUT
(Unit Cost Index Numbers, 1929 = 100)

Item	1870– 1900	1900	1910	1919	1929	1937	1948	1957
Gross national product, less extractive goods	136	126	115	118	100	102	80	68
Agriculture	132	118	121	114	100	93	73	66
Minerals	211	195	185	164	100	80	61	47
Saw logs	59	65	67	108	100	104	88	90
Agriculture, relative to GNP	97	94	105	97	100	91	91	97
Minerals, relative to GNP	155	155	161	139	100	78	76	69
Saw logs, relative to GNP	36	47	55	86	100	101	106	130

SOURCE: H. J. Barnett and Chandler Morse, *Scarcity and Growth* (Baltimore: Johns Hopkins Press, 1963), pp. 205–7.

It is against this background of rapidly rising demand, countered in the developed economies by an even more rapid increase in man's ability to rearrange productive processes and even the composition of the earth itself, that the role of marine mineral resources must be assessed.

To Supply America's Needs

It would now appear, on the basis of the best professional evidence at hand, that supplies of minerals available to the United States from conventional land sources are adequate to meet projected demands to 1985 and 2000 without significant increase in prices. This is certainly true for some mineral commodities, including coal, iron, magnesium, bromide, lithium, phosphate, potassium, and sodium, and less certainly for copper, molybdenum, and sulphur. For some others, such as chromium, manganese, nickel, cobalt, industrial diamonds, platinum, and tin, the same answer could be given on a world basis; but the United States is, or will become, almost completely dependent on foreign sources of supply.

When we consider that minerals make up only a small fraction of gross national product at the present time, it is evident that increases in the costs of minerals are not going to exert any major influence on American economic growth (or on world economic growth), at least over the next thirty years or so. Thus, minerals from the sea will become important to us not as a God-given respite from a process otherwise leading to economic disaster, but rather as a welcome source of lower-cost raw materials—surely a more comfortable prospect.

In summary, currently available data indicate that with very few exceptions (notably gold, silver, and uranium) the world supply of hard-mineral resources from conventional sources is sufficient to meet projected demands at or close to current relative prices at least to the year 2000. It should be stressed that this judgment is based on the assumption that present rates of technological progress will continue, and that new discoveries on land will proceed, sporadically, at a rate comparable to that of recent decades. Equally important, it must be assumed that international military and political considerations will still permit the Free World to maintain or expand the level of international trade, and that the United States will therefore continue to have access, at favorable prices, to the important list of minerals not fully available from domestic sources.

The importance of the latter assumption is critical, and failure to explore its full ramifications has resulted in wide differences of opinion as to appropriate national policy. It is justifiable, therefore, to digress for a moment to consider the alternatives that might face the United States if political or military considerations should block American access to the lowest-cost sources of strategically important minerals. First, it is possible, at reasonable cost, to stockpile most of the items that might be considered of critical importance in various industrial processes. Many of these are required only in minute quantities, and a stockpiling program would not involve massive commitments. Second, as is true for almost any other natural resource, there will always be marginal and submarginal suppliers, at any given price level, whose output could be expanded in times of disruption by the process of offering greater economic incentive. There are "industries in being," some domestic and some in nonsensitive political areas, that could be drawn upon, albeit at higher cost, if normal supplies were interrupted. Third, the quantity of mineral input required per ton of a given output is not a fixed quantity. Thus, process changes and substitutions would make it possible, in case of real emergency, to reduce total demand significantly without serious impacts on national output, at least in the short or intermediate term.

Note that all of these optional alternatives to present sources of supply of imported mineral products are also alternatives to the development of marine-based sources. Their availability strengthens the position that the mineral resources of the sea should not be viewed as a "last gasp" source of critical inputs, to be developed regardless of cost.

Types of Marine Mineral Resources

In a purely physical sense, marine mineral resources can be divided into four major types, based on the environment in which they may be found: (1) dissolved chemicals in sea water; (2) submerged placer deposits that occur as patches of unconsolidated material on and immediately under the continental shelf; (3) deposits in the substrate rocks of the continental margin (and perhaps of the deep ocean basins); and (4) deposits on the abyssal plains of the deep ocean basins.

Despite the fun and games that seem to be popular in amateur oceanographic circles at the moment, involving the calculation of total tonnage of dissolved minerals in the world's oceans, the chemical constituents of sea water do not appear to offer much economic promise beyond the considerable production already under way. The sea is a major source of bromine, magnesium, and salt. Indeed, these materials are present everywhere in the sea in sufficient concentration that they may be considered inexhaustible. Production is therefore limited by market demand, and costs of producing these materials from sea water set an effective ceiling on prices that could be obtained from any other source.

For the rest, however, the sea provides a lean ore indeed. To illustrate, the gross value per cubic mile of sea water of seventeen common industrial elements of commercial significance would be slightly less than $600,000 at current prices. Yet a plant capable of recovering the seventeen elements would have to be large enough to pump about 2 million gallons per minute (Commission on Marine Science, Engineering and Resources 1969, p. VII–101). The cost of an operation of this kind would be far greater than the gross value of the minerals recovered even if some concentration were achieved prior to processing (as from the outfall of a plant producing fresh water from sea water). Expressed in other terms, a barrel of sea water contains enough table salt to provide for about two years of normal consumption by one person, enough magnesium to make a half-pound ingot, and enough bromine to provide antiknock additive for about ten gallons of gasoline. But to get seventy-five cents' worth of gold would require nearly a million drums of sea water. The sea's contribution to United States supplies of sodium chloride, bromine, and magnesium is welcome; for the rest, concentrations are so thin as to suggest that more attractive sources are to be found on land.

Submerged placer deposits provide the most promising source of hard minerals from the seabed. As geological extensions of mineral sources on land, these offer much less imposing obstacles to discovery and recovery than do hard minerals bedded in the substrate of the sea bottom. The gold of Alaska, the Cornish and Malaysian tin mines, and the diamond operations of southwest Africa are well-known examples of marine placer operations which are really extensions of land-based activity. As indicated in Table 6, more mundane products—sand, gravel, and oyster shells—are far more important in total value at present; indeed, practically all of the mineral output from the sea by United States firms is now concentrated in these three products.

TABLE 6

OFFSHORE MINING OPERATIONS IN 1967:
UNCONSOLIDATED DEPOSITS

Mineral	Location	Number of Operations	Annual Production	Year	Value (millions of dollars)
Diamonds	Southwest Africa	1	221,500 cubic yards	1964 1965	$ 8.9
Gold	Alaska	1	. . .	1966	. . .
Heavy mineral sands	America, Europe, Southeast Asia, Australia	15	1,307,000 tons	1965	13.1
Iron sands	Japan	3	36,000 tons	1962	3.6
Tin sands	Southeast Asia, U.K.	4	10,000 tons concentrate	1965	24.2
Lime shells	U.S.A., Iceland	9	20,000,000 cubic yards	1965	30.0
Sand and gravel	U.K., U.S.A.	38	100,000,000 cubic yards	1966	100.0
Total		71			$179.8

SOURCE: Charles M. Romanowitz, Michael J. Cruickshank, and Milton P. Overall, "Offshore Mining Present and Future," presented at NSIA/OSTAC Ocean Resources Subcommittee Meeting, San Francisco area, 26 April 1967.

Table 7 lists some of the placer minerals regarded as good prospects for development on the continental shelves on North America. None is currently in production, but all possess enough potential so that continued improvement in discovery and recovery techniques would move them into the cost-effective range.

In somewhat the same category, deposits of marine phosphorite nodules are known to exist on the continental margins. They are not of the same origin as the placers, but apparently are derived from erosion of exposed phosphatic beds or are precipitated from sea water, forming in areas of cold upwelling, rich in nutrients.

The most valuable resources to be found in the substrate rocks of the continental shelves and slopes are petroleum and gas, but there are also sub-

TABLE 7
UNITED STATES WEST AND EAST COAST
PLACER MINERALS

West Coast	East Coast	West Coast	East Coast
Gold	Ilmenite	Ilmenite	Zircon
Platinum	Rutile	Zircon	Kyanite
Cassiterite	Monazite	Rutile	Sillimanite
Magnetite	Xenotime	Chromite	Staurolite

SOURCE: Commission on Marine Science, Engineering and Resources 1969, p. VII–101.

stantial and profitable operations in the production of sulphur, confined at present to the Gulf of Mexico. The rate of increase in sulphur production from the salt domes in the Gulf of Mexico is expected to taper off, however. Thus far, the only other operations involving offshore mining of consolidated products are the Finnish and Newfoundland iron ore operations and the production of coal in Nova Scotia, Taiwan, Japan, Turkey, and the United Kingdom. There is no reason to doubt that other potentially important resources lie bedded in the rocks of the shelves and slopes. These include iron ore, bauxite, coal, phosphatic rock, tin, and bedded salt that may contain potash. Onshore lode and bedrock deposits are sufficiently abundant near the coasts of Alaska, the other West Coast states, and New England to suggest the possibility of offshore deposits with the same mineral characteristics.

One must conclude with the obvious (but important) statement that knowledge of the hard-mineral potential of the continental shelves and slopes is so slight that generalizations are virtually impossible. Until such time as broad-grid bathymetric mapping and charting (with much finer analysis of promising areas) and geological analysis are far more advanced, no quantitative appraisal can be formulated that carries much physical or economic meaning. Obviously, the same comments apply with even greater force to mineral deposits on and under the deep ocean floors. At the moment, the now-famous manganese nodules are the only known minerals of potential economic importance at these depths, although other materials, principally chromite and nickel, may well be found in some areas.

The technical capacity to explore and exploit profitably the minerals on the continental shelves is so limited that hard-rock mining in the ocean deeps hardly warrants comment at this point. Nevertheless, the manganese nodules (which contain significant amounts of copper, cobalt, and nickel in addition to manganese oxide) have achieved so much prominence that some discussion of their potential seems warranted.

The fact that the nodules are on the continental slopes and deep ocean floor in quantities that stagger the imagination has been established beyond question. But analysis of the significance of their existence for the economic activities of man suggests a much more sober appraisal. The known locations of commercially interesting concentrations lie at great depths: recovery would have to be on a very large scale if the "fixed cost" of technological development required to exploit the nodules were to be held to tolerable levels per unit of output. One expert has estimated that one optimal-sized operation would suffice to meet 25 percent or more of all United States requirements (Brooks 1966).

Since manganese accounts for a very small part of the total cost of the end products for which it is used, demand is likely to be highly inelastic. Thus, a sudden increase in United States output ranging from 25 to 50 percent (the minimum economic operating level for one or two operations) would have a major depressing effect on prices. The chemical and physical characteristics of the nodules are such that specialized processing facilities would have to be developed and constructed. Finally, total annual consumption of manganese, relative to known world supplies available at costs well below even the most optimistic estimates for offshore sources, suggests that the sea-floor nodules could not compete on a straight commercial basis for decades to come.

If, on the other hand, manganese nodules are regarded as an emergency supply, to be utilized if political or military developments bar access to cheaper land-based sources overseas, subsidized development must be weighed against alternative methods of achieving the same result. The physical requirements of American industry are small enough so that stockpiling becomes a reasonably economic alternative way of assuring continued supply until longer-term adjustments to the new situation could be completed. The same argument would apply to the development of high-cost domestic land-based supplies. The world is, after all, full of natural resources that are unutilized and may remain unutilized for centuries simply because their recovery would not pay. Manganese nodules may fall in this category, at least for the near future.

On the other hand, it would be irresponsible to ignore the impact of continuous but often unpredictable technological change on the dividing line between economically feasible and unfeasible operations. Manganese nodules can properly be regarded as establishing a ceiling on possible price increases in materials from competitive land-based sources. It is entirely possible, however, that a major national effort to improve basic technology for operation in the sea, coupled with specific development projects for recovery and handling of underseas minerals, could lower that ceiling much more rapidly than is now anticipated.

TABLE 8

VALUE OF MINERAL PRODUCTION FROM OCEANS BORDERING THE UNITED STATES
1960–67
(Millions of Dollars)

Commodity	1960	1961	1962	1963	1964	1965	1966	1967
From sea water:								
Magnesium metal and compounds, salt, and bromine	69.0	73.0	89.1	84.6	94.5	102.6	177.0	145.4
From ocean subfloors:								
Petroleum, natural gas, and sulphur	423.6	496.6	620.7	730.8	820.3	933.3	1177.7	1404.8
From beaches and inshore sea floors:								
Sand and gravel, zircon, feldspar, cement rock, and limestone	46.8	46.2	44.3	42.5	43.6	51.4	51.6	55.9
Total	539.4	615.8	754.1	857.9	958.4	1087.3	1406.3	1606.1

SOURCE: Commission on Marine Science, Engineering and Resources 1969, p. VII–101.

Table 8 provides a summary of current hard-mineral production from marine sources adjoining the United States. As of 1967, less than three hundred active operations were engaged in production of hard minerals from marine sources throughout the world. The total value of their output was about $712 million, of which a major part entailed recovery of dissolved minerals (salt, magnesium, and bromine), sand and gravel, and sulphur. Offshore operations in Thailand and Indonesia account for more than 10 percent of the world's tin production, but are carried on only in very shallow waters. Despite these small numbers, however, the level of exploration and technological inquiry by private mining companies and by national government foretells a far larger future level of operation.

The Role of Government

The future of marine hard minerals in the United States cannot be viewed as a resultant of market forces alone. Very expensive exploration, geological analysis, and broad-spectrum technological development must be undertaken before major increases in production can be expected. Yet the benefits that accrue from these essential activities cannot be captured by single firms or groups of firms or even, in some cases, by single nations. Thus, the speed with which production functions for marine minerals are shifted to produce economically feasible cost levels is a matter of investment policy by both government and private enterprise; and the role of government must be conditioned not only by economic objectives but also by considerations of national security and regional economic development. Moreover, the kind of exploration and developmental work required of government will almost certainly have beneficial economic effects that extend far beyond the mining industry alone to become, of necessity, a part of the general program for advancement of scientific knowledge of the seas.

These latter considerations seem to be the governing motivations for an expanded national effort directed toward three-dimensional exploration of the continental shelves (and, later, of the slopes) and toward the basic technology required to work effectively in the sea. While one output of a programmed and systematic effort along these lines may be to push forward the date at which marine mineral resources can become competitive, there is no reason to view this as a major objective in its own right. In short, if more rapid and broader utilization of minerals from the marine environment occurs as a result of a balanced national effort to learn more of the sea and its potential contribution to man's welfare, so be it. But to set out, deliberately, to expand the proportion of total resource consumption from marine resources regardless of relative cost would involve a major commitment of government resources to projects with a return, by definition, lower than could be achieved in other applications. From the economic point of view, it would appear much more sensible to thank our lucky stars for the opportunity to approach the mineral resources of the sea as part of a broad spectrum of public and private inquiry and with sufficient time to avoid the inevitable waste of crash programs.

Space limitations preclude any detailed examination of the policy alternatives that should be considered in developing national policies toward the sea and its resources. As the discussion above suggests, the problems and remedies involved go far beyond the confines of mineral extraction as such, since the barriers to more rapid development are as much scientific and institutional as they are narrowly technological. For a full range of the complex issues involved, the reader is referred to the report of the Commission on Marine Science, Engineering and Resources, and the more detailed reports of its several panels. In addition, the report of the Public Land Law Review Commission contains important current information on the status and prospects for mineral activity on the continental shelf.

CONCLUSIONS

It is apparent that the role to be played by the marine environment in the broad context of world and national resource demand and supply will grow in importance. Even the fisheries, which suffer almost as much from overdevelopment as from underdevelopment, are capable of making a much more significant contribution to the never-ending fight against hunger and malnutrition than present knowledge and institutions permit. Petroleum, natural gas, sulphur, and the three principal dissolved materials now being extracted commercially already make up a significant part of total world output. All will assume greater importance in the foreseeable future. The hard-mineral field is hardly past the talking stage as yet, but it would be incredible if the pace of man's knowledge of his environment and how to use it, even without the stimulus of a new national program, failed to bring these potential assets into economic utilization before too many decades pass.

But notice the stress of the word "potential." Our ability to convert the marine environment to beneficial uses is science-limited, technology-limited, and institution-limited. In the case of both fisheries and petroleum resources, the gaps in scientific knowledge and the technological needs are very real; but the more immediate impediments to progress lie in our apparent inability to devise legal, institutional, and administrative arrangements that will permit optimal rates of exploitation and an equitable division of both end products and employment opportunities among participating states and nations. As Professor Auerbach's paper suggests, even the production of hard minerals (most of it still in the distant future) will continue to be hampered by the confused state of the legal rights of explorers and investors in mineral resources on or beneath the seabed. It is particularly frustrating for social scientists to acknowledge that man's scientific and technical capabilities are not matched by an equal ability to arrange for their effectve conversion into human welfare.

The final conclusion is foreshadowed by the discussion above. The level of effort—scientific, organizational, and technological—devoted to study of the ocean environment and its resources should be expanded sharply. But it would be logically contradictory and administratively impossible to separate out for special consideration programs aimed only at speeding the development of marine resources. We must advance simultaneously on a broad front, with as much emphasis on basic science and technology as on specific mission-oriented tasks, if the ocean's resource potential is to become reality.

REFERENCES

Brooks, David
  1966   *Low-Grade and Non-Conventional Sources of Manganese.* Washington, D.C.: Resources for the Future.

Christy, Francis, Jr., and Anthony D. Scott
  1965 *The Common Wealth in Ocean Fisheries.* Baltimore: Johns Hopkins Press.
Commission on Marine Science, Engineering and Resources
  1969 *Marine Resources and Legal-Political Arrangements for Their Development.* Panel Reports, III. Washington, D.C.: U.S. Government Printing Office.
Davidson, Paul
  1963 "Policy Problems of the Crude Oil Industry." *American Economic Review* (March), pp. 85–108.
Lovejoy, W., and P. Homan
  1967 *Economic Aspects of Petroleum Conservation Regulation.* Baltimore: Johns Hopkins Press.
McKie, James, and Stephen MacDonald
  1962 "Petroleum Conservation in Theory and Practice." *Quarterly Journal of Economics* (February), pp. 98–121.
Schaefer, Milner B., and Dayton L. Alverson
  1968 "World Fish Potentials." In De Witt Gilbert, ed., *The Future of the Fishing Industry of the United States.* Seattle: University of Washington. Pp. 81–101.
Taylor, Donald M.
  1969 "New Production Ideas and Concepts." *Ocean Industry* 4, no. 1, pp. 27–32.

An International Legal-Political Framework for Exploring and Exploiting the Mineral Resources Underlying the High Seas: The Recommendations of the Commission on Marine Science, Engineering and Resources

CARL A. AUERBACH

SINCE my meeting with the seminar in May 1968, the report of the Commission on Marine Science, Engineering and Resources (hereafter referred to as COMSER Report) has been submitted to the president and Congress. It is more important, therefore, to present COMSER's conclusions than the recommendations in the report of its International Panel (hereafter referred to as International Panel Report) which were discussed at the seminar. On the whole, I should hasten to add, COMSER adopted these recommendations (see International Panel Report).

During the period of COMSER's existence, much greater interest was manifested in and out of government in the international framework for subsea mineral resource exploration and development than that for the exploitation of high seas fisheries. Yet the annual value of the world catch of fish and shellfish is nearly one and one-third times that of all other marine resources (COMSER Report, p. 11). This emphasis on minerals may be explained, however, by the greater glamour in the prospect of extracting them from the seabed and by the new legal problems that will have to be solved to convert this prospect into reality. Considerable riches are also at stake, involving the strong and articulate petroleum and mining industries in the United States and abroad.

Carl A. Auerbach is professor of law at the University of Minnesota.

OBJECTIVES OF FRAMEWORK

In recommending a framework—the principles and rules, institutions and procedures—for the exploration and exploitation of subsea mineral resources, COMSER sought to attain the following objectives: (1) to promote international peace and order; (2) to give the United States and all other nations an equal chance to develop and to benefit from the exploitation of mineral resources; (3) to encourage scientific and technological efforts and the necessary major capital investments by making it possible to conduct these activities in an orderly and economic manner; (4) to minimize the creation of vested interests which might inhibit changes in the framework that will be deemed desirable in the light of unfolding experience with actual exploration and exploitation (COMSER Report, pp. 141, 143).

To achieve these objectives, COMSER concluded, the framework must provide the incentive to undertake the exploration and exploitation of mineral resources by recognizing exclusive claims to large enough subsea areas for long enough periods of time. It must also protect recognized claims and at the same time require the relinquishment of claims that are not properly explored or developed within fixed reasonable periods of time. Finally, it must provide for the peaceful settlement of any disputes that might arise (COMSER Report, p. 143).

The existing international framework does not provide the necessary means to achieve these objectives. General principles of international law recognize the right of each coastal nation to permanent, exclusive access to the mineral resources found either in its territorial waters, or on their beds or in their subsoil. But there is no international agreement on the breadth of the territorial sea, and claims vary from three to more than two hundred nautical miles (see International Panel Report, Chap. 8, p. 11).

The International Convention on the Continental Shelf * grants to each coastal nation "sovereign rights" over the continental shelf "for the purpose of exploring it and exploiting its natural resources" (Art. 2 [1]). But the definition of the "continental shelf" imposes uncertain limits, if any, upon its extent. Assuming there are such limits, only general principles of international law govern exploration and exploitation of the mineral resources of the bed and subsoil of the subsea areas beyond them. And these principles themselves abound with uncertainty.

DEFINITION OF "CONTINENTAL SHELF"

Without exception, United States private enterprise made it clear to COMSER that it would not proceed to explore and exploit the mineral re-

* International Convention on the Continental Shelf, adopted by the United Nations Conference on the Law of the Sea, 29 April 1958 (hereafter International Convention). The convention entered into force in 1964. 15 U.S.T. 471, T.I.A.S. No. 5578, U.N. Doc. No. A/CONF. 13/L.55 (1958).

sources of the bed and subsoil underlying the high seas unless it was assured of exclusive access to such resources in large enough areas for long enough periods of time to make the activity profitable. Yet no one can reasonably say that the existing framework assures this security much beyond the two-hundred-meter isobath. The principal uncertainty derives from the convention's definition of the continental shelf, which extends the shelf "to the seabed and the subsoil of the submarine areas adjacent to the coast but outside the area of the territorial sea, to a depth of 200 meters or, beyond that limit, to where the depth of the superjacent waters admits of the exploitation of the natural resources of the said areas . . ." (International Convention, Art. 1). It should be noted that this legal definition of the shelf does not correspond to its geological definition.

Even the coastal nation's right of permanent, exclusive access to the natural resources of the continental shelf up to the two-hundred-meter isobath is not entirely free from doubt, because in some parts of the world the two-hundred-meter depth is so far from the coast that at some point, it may reasonably be argued, it is no longer "adjacent" to it and therefore is not within the convention's definition. As a practical matter, however, there seems to be general international recognition of the coastal nation's rights up to the two-hundred-meter isobath.

At the other extreme, it has also been argued that as soon as it becomes possible technologically to exploit mineral resources to any ocean depth, the seabed and subsoil of all the submarine areas of the world, totaling 139 million square miles, will become parts of the continental shelves of coastal nations (see, for example, Bernfeld 1967, pp. 72–73; Oda 1967, pp. 6–7). These nations would then divide the areas among themselves in accordance with Article 6 of the International Convention on the Continental Shelf. In effect, this article would then give each coastal nation sovereign rights to the natural resources of the seabed and subsoil from its coasts to the median line, that is, the line on which every point is equidistant from the coasts of the nearest nations.

It may now be said that this reading of the definition of the continental shelf has been repudiated by the nations of the world. The report of the United Nations ad hoc committee, whose work will be considered at some greater length below, "recognized the existence of an area of the seabed and the ocean floor underlying the high seas beyond the limits of national jurisdiction." * This recognition, of course, is inconsistent with the "median line" concept, the adoption of which would have deprived the United Nations ad hoc committee of its reason for existence. Though it rejected the median-line concept, the committee made no effort to define the "limits of

* Report of the Ad Hoc Committee to Study the Peaceful Uses of the Sea-Bed and the Ocean Floor beyond the Limits of National Jurisdiction, United Nations General Assembly, Official Records, Twenty-Third Session, UNGA Doc. A/7230 (1968).

national jurisdiction," that is, the limits of the continental shelf as defined in the International Convention.

How far beyond the two-hundred-meter isobath does the legal continental shelf extend? The view has been expressed that the exploitability criterion in the definition of the continental shelf was intended only to permit a coastal nation to extend beyond the two-hundred-meter isobath particular operations which it began on the continental shelf at depths less than two hundred meters (Goldie 1968, p. 8). The National Petroleum Council (NPC), however, takes a far more expansive view of the exploitability criterion. It maintains that the exploitability and adjacency criteria, taken together, give coastal nations "sovereign rights" over the natural resources of the continental land mass seaward to where the submerged portion of that land mass meets the abyssal ocean floor. This view would give the coastal nations permanent, exclusive access to the natural resources of the geological continental shelves, continental borderlands, continental slopes, and at least the landward portions of the geological continental rises. Where the continent drops off sharply from near the present coastline to the abyssal ocean floor, the NPC would add to the legal continental shelf an area of that floor contiguous to the continent.*

The NPC recognizes that its interpretation of the continental shelf definition is not the only reasonable one and that United States private enterprise will be hesitant to invest large amounts of capital in subsea areas of uncertain status. To remove the uncertainty, it recommends that the United States declare its acceptance of the NPC interpretation and its intention to exercise its "sovereign rights" accordingly and invite all other coastal nations to make similar declarations (NPC Interim Report, p. 6).

The NPC position has received support elsewhere. At its annual meeting in August 1968, the House of Delegates of the American Bar Association (ABA) passed a resolution stating that "it is generally recognized that the definition in the 1958 Convention on the Continental Shelf of the boundary between the area of exclusive sovereign rights and the deep ocean floor needs to be clarified by an agreed interpretation." But it recommended only that "the United States consult with other parties to the 1958 Continental Shelf Convention with a view to establishing, through the issuance of parallel declarations or by other means, an agreed interpretation of the definition of the boundary between the area of exclusive sovereign rights with respect

* *Petroleum Resources under the Ocean Floor: An Interim Report by the National Petroleum Council's Committee on Petroleum Resources under the Ocean Floor* (hereafter NPC Interim Report), 9 July 1968, pp. 9, 15. The interim report was adopted by the NPC on 9 July 1968, with a note that "nothing contained" therein "should be construed as representing U.S. policy." The final NPC report, approved in April 1969, affirms the conclusions of the interim report. See *Petroleum Resources under the Ocean Floor: A Report of the National Petroleum Council* (1969).

to natural resources of the seabed and subsoil and the deep ocean floor beyond the limits of national jurisdiction." * The resolution, however, did not
propose "an agreed interpretation" of this boundary.

The report of the American Bar Association Sections of Natural Resources Law and International and Comparative Law and the American Bar
Association Standing Committee on Peace and Law through United Nations, which accompanied the resolution passed by the association's House
of Delegates, expressed the opinion that the "better view" of the definition
of the legal continental shelf was that advanced by the NPC. The report also
endorsed the NPC position that existing legal uncertainties should be elimi-
nated by "uniform declarations of the coastal nations which are parties to
the Convention on the Continental Shelf, identifying their claims of jurisdiction with the submerged portion of the continental land mass." †

It should be pointed out that, under ABA practice, approval of a resolution by the House of Delegates does not imply approval of the accompanying report. Furthermore, representatives of the ABA groups assured me
that they would reconsider their position in the light of the reports of
COMSER and its International Panel. The Committee on Deep Sea Mineral
Resources of the American branch of the International Law Association (ILA)
has also allied itself with the NPC stand.‡

There was an overlapping of memberships in the NPC, ABA, and ILA
groups mentioned above. No one person served on the committees of all
three organizations; but three individuals did serve on both the NPC and
ILA committees; two on those of the ABA and ILA, and one on the NPC
and ABA committees. More important, the NPC, ABA, and ILA groups
relied heavily on a very summary review, prepared by Oliver L. Stone, general attorney of the Shell Oil Company, of the history of the 1958 Geneva
Convention on the Continental Shelf. Stone was a member of both the NPC
and ABA groups, and his memorandum was reproduced as an appendix to
both the NPC and ILA interim reports.

On 6 October 1968 the board of directors of the American Mining Congress adopted a "Resolution on Undersea Mineral Resources" which stated
that the "boundary between the area which by treaty is now under national
jurisdiction (the 'continental shelf' of the 1958 Convention) and the area beyond (the deep ocean) should be defined with precision and with due regard
to the preservation of the national interests to the full extent permitted by

---

* See *Proceedings of House of Delegates, American Bar Association,* Philadelphia,
Penn., 5–8 August 1968, 54 A.B.A.J. 1017, 1030 (1968).
† See Joint Report on Submarine Mineral Resources, 1968, A.B.A. Rep.
‡ Interim Report of Committee on Deep Sea Mineral Resources of the American
Branch of the International Law Association, 19 July 1968 (hereafter ILA American
Branch Interim Report), pp. ix–xii. Professor Louis Henkin of the Columbia Univer-
sity Law School dissented from the report, expressing opinions shared by COMSER
and its International Panel. Ibid., p. xxi.

the 1958 Convention on the Continental Shelf." But the resolution did not specify how such a boundary should be defined. Speaking for the American Mining Congress at a government-industry meeting on 21 October 1968, Malcolm R. Wilkey, general counsel and secretary of the Kennecott Copper Corporation, explained this resolution.* It appears that he disagreed with the NPC position because it called upon the United States to stake out an immediate claim to the mineral resources of the subsea areas up to the base of the continental slope even though such depth was not yet exploitable. Wilkey, however, thought that the 1958 convention would permit the staking of such a claim when technology actually carried exploitability to the foot of the continental slope. At the same time, he advanced cogent reasons against the advisability of the resulting wide continental shelf.

COMSER's International Panel studied the history of the 1958 Geneva Convention and the work of the International Law Commission preparatory to it. At its request, too, Bernard H. Oxman, then Lieutenant, USNR, and assistant head of the Law of the Sea Branch, International Division, Office of the Judge Advocate General, made an independent study of the same history (Oxman 1968). These studies reveal isolated bits and pieces of "legislative history" which offer some support for the position of Stone and the NPC. As a good lawyer trying to prove a point, Stone has assembled these bits and pieces in his memorandum. But neither the language of the definition of the continental shelf nor its history as a whole supports the NPC position. There is stronger evidence for alternative interpretations.

The fact is that the draftsmen of the convention did not anticipate that advancing technology would give the exploitability criterion the pivotal importance it has assumed. The convention intended, primarily, to codify the principles on which the Truman Proclamation of 1945 was based.† But the Truman Proclamation asserted sovereign rights over the natural resources of the continental shelf only as geologically defined. Even though the continental shelf's legal definition does not coincide with its geological definition, it was not intended to abandon entirely the link to the geological shelf.

The view of adjacency upon which the NPC position is based is not that which was held by the International Law Commission. The ILC intended this criterion to place a limitation on the exploitability criterion which, in turn, was intended to give "sovereign rights" to coastal nations in waters less than two hundred meters deep that were not on a geological shelf, as in the Persian Gulf, as well as to nations, particularly in South America, whose coasts dropped almost immediately to great depths and who insisted

* The minutes of this meeting were made available to COMSER and to the author by Dr. Edward Wenk, Jr., at that time executive secretary of the National Council on Marine Resources and Engineering Development.

† Presidential Proclamation No. 2667, 28 September 1945, "Policy of the United States with Respect to the Natural Resources of the Subsoil and Seabed of the Continental Shelf," 10 Fed. Reg. 12303 (1945).

on getting equal rights under the convention. The International Panel agreed with the conclusion of Louis Henkin that there is no way "to 'redefine' the outer limit" of the continental shelf "by interpretation of the Convention" (Henkin 1968, p. 21). This conclusion was also accepted by COMSER (COMSER Report, p. 144).

At first encounter, the NPC proposal to eliminate uncertainty in the continental shelf definition by having the president of the United States issue another Truman-like proclamation seems attractive. Such a proclamation along NPC lines—which the United States has the power to effectuate—would add 479,000 square statute miles of seabed and subsoil to the approximately 850,000 square statute miles of continental shelf up to the two-hundred-meter isobath. It is estimated that 181 billion barrels of petroleum and 1,440 billion cubic feet of gas may ultimately be produced from this additional vast area.* The United States could make itself the sole beneficiary of these rich deposits. Furthermore, some United States oil companies seemingly would prefer to continue to face the known perils of the exercise of exclusive authority by coastal nations around the world than the unknown perils of international legal-political arrangements yet to be negotiated.

But after due consideration, COMSER rejected the NPC proposal because it is contrary to the best interests of the United States. In the first place, it would benefit other coastal nations of the world proportionately more than the United States and would give them permanent, exclusive access to the natural resources of immense subsea areas.† In light of recent history, it is shortsighted to assume that United States private enterprise would be better off to deal with these coastal nations for permits to develop these resources in the absence of any recognition of the interest of the international community in them.

Secondly, some other coastal nations will not benefit even as much as the United States from the NPC proposal. Francis Christy estimates that fifty to sixty of the coastal nations would benefit very little therefrom because they have small toe holds on the oceans, are situated on small seas, or confront islands belonging to other sovereign nations lying between them and the open seas; and another twenty-five to thirty would benefit only moderately. But twenty-five to thirty coastal nations would benefit greatly (Alexander

* Report of Panel on Marine Resources of Commission on Marine Science, Engineering and Resources (Washington, D.C.: U.S. Government Printing Office, 1969).

† The 479,000 and 850,000 square statute miles figures given above are based upon estimates of the U.S. Geological Survey. The equivalent figures in square nautical miles are approximately 361,000 and 630,000, respectively. Around the world, the number of square nautical miles between the 200-meter and the 2,500-meter isobaths is approximately equal to the number of square nautical miles of continental shelf up to the 200-meter isobath.

1969). The NPC proposal would create the danger that the coastal nations which did not benefit appreciably from it for any of the reasons mentioned by Christy, or because there were no important mineral deposits in the extended submarine areas allotted to them, might feel justified in claiming as compensation exclusive access to the superjacent waters, the living resources in them, and the air above them.

The danger that rights of permanent exclusive access for one purpose may expand to claims of exclusive access for other purposes or even of territorial sovereignty (that is, permanent, exclusive access for all purposes) materialized as an unforeseen and undesirable consequence of the Truman Proclamation of 1945. It materialized even though the proclamation, like the Convention on the Continental Shelf (Article 3), made it clear that the exercise of sovereign rights over the natural resources of the continental shelf was not to affect the legal status of the superjacent waters as high seas, or that of the air space above those waters.

The unilateral reactions of Chile, Ecuador, and Peru to the Truman Proclamation culminated in the Declaration of Santiago, on 18 August 1952, in which the three countries proclaimed "as a principle of their international maritime policy that each of them possesses sole sovereignty and jurisdiction over the areas of the sea adjacent to the coast of its own country and extending not less than two hundred nautical miles from the said coast." While Chile and Peru subsequently made it clear that they claim only exclusive fisheries zones of not less than two hundred miles, Ecuador claims a territorial sea of two hundred miles and has recently issued a decree that appears to attach restrictions upon the right of foreign naval vessels and aircraft to transit or fly over its claimed ocean area.

These claims have harmed our distant-water fishing fleet and our general relations with these countries (see International Panel Report, p. VIII-23). In the case of Peru, the resulting difficulties may jeopardize the chance of settling our oil disputes with that country. The very industry that the Truman Proclamation was intended to help may thus have been hurt.

It has been traditional United States policy to limit national claims to the sea in the interest of the maximum freedom essential to the multiple uses, including military uses, which the United States makes of the oceans. Indeed, it may be asserted with confidence that the president of the United States would never have signed, and the Senate of the United States never would have ratified, the Convention on the Continental Shelf if they thought it embodied the position the NPC now takes. For the national security and world peace are best served by the narrowest possible definition of the continental shelf for purposes of mineral resource development. The NPC proposal is also unfair to the land-locked nations of the world which will want to know why the rich mineral deposits on and under the continental slopes and rises should belong only to coastal nations. United States initiatives to implement the NPC proposal would be regarded as a "grab," even if other coastal nations followed suit.

For these reasons, COMSER recommended that the United States lead an effort to reach international agreement upon a precisely redefined "narrow" continental shelf (COMSER Report, pp. 145–46). Specifically, it recommended that the seaward limit of each coastal nation's "continental shelf" should be fixed at the two-hundred-meter isobath or at fifty nautical miles from the baselines for measuring the breadth of its territorial sea, whichever alternative gave it the greater area for purposes of the Convention on the Continental Shelf. If the same continental shelf, as redefined, is claimed by two or more nations whose coasts are opposite each other, or by two or more adjacent nations, COMSER recommended that the boundaries should be determined by applying the median-line principles set forth in Article 6 of the convention.

By providing the depth and distance-from-shore alternatives, COMSER sought to avoid the inequity of a depth criterion alone for those coastal nations which either are not on a geological continental shelf, as in the Persian Gulf, or have coasts that drop to great depths almost immediately, as off the west coast of South America. The two-hundred-meter/fifty-mile pairing was chosen because it is about as close together as pairings on world-wide averages of the depths and widths of the world's geological continental shelves can reasonably be.

SUBSEA AREAS BEYOND THE CONTINENTAL SHELF

Reasonable arguments have been advanced for and against the proposition that general principles of international law support one or more of the following views. (1) Any nation may acquire "territorial sovereignty," which is the right of permanent, exclusive access for all purposes over any area of the seabed and subsoil beyond the limits of the continental shelf which that nation is the first to occupy. (2) Any nation may acquire "sovereign rights," which grant permanent, exclusive access to the mineral resources of any such area which the nation is the first to discover and exploit. (3) No nation may claim such "sovereignty" or "sovereign rights," but any nation may extract and keep the mineral resources from any such area and protect the operations which carry its flag. (4) No nation may extract mineral resources from any such area without the permission of the international community.

The report of the ABA groups states that there "appears to be general agreement, both in and out of government here and abroad" that position 1 above is untenable. It adds that there "is also general agreement that a nation which undertakes the exploration and exploitation of mineral resources on and under the deep seabed should be protected in the exclusive right to occupy the areas involved, with due regard to other uses of the marine environment, and without impairment of the high-seas character of the overlying waters." * While this statement is not wholly free from ambiguity, it seems to assume general acceptance of position 2. With all due respect, it must be said that there is not general acceptance of the view that sovereign rights may be acquired beyond the limits of the continental shelf.

There is strong support for position 3; but there is no agreement that a nation which is the first to develop the mineral resources of such a subsea area may exclude "poachers" or other operators who wait until the operator carrying its flag has done the exploration and made the discovery and then, having avoided the costs of exploration, move in to extract minerals from the same area. Neither the NPC nor the ILA interim report supports the ABA report in this respect.

The American Mining Congress Resolution on Undersea Mineral Resources asserted that general legal principles "already govern ocean activity but certain rights need sharper definition, for example, in regard to the right to explore, the right to access, the exclusive right to mine if an ore body is found, and suitable space for ancillary purposes." Wilkey explained this resolution as follows:

> We take the position that there is no legal void in the deep ocean, that the long-established general principles of international law apply, that these principles are sufficient to protect American industry representatives in exploring the subsoil of the deep ocean, but we recognize that these general principles need considerably more specificity in order to protect the detailed rights needed for large expensive mineral development. In other words, the principles of freedom of the seas will be sufficient for the next few years of exploration and preliminary development, the American State Department and Navy should offer diplomatic and naval protection for American industry efforts, but the legal principles of freedom of the seas were never designed with sufficient specificity to protect mining rights on the ocean floor, and therefore more detailed rules will be needed in the future.

Essentially, the American Mining Congress and Wilkey also agree with COMSER that the framework for the development of subsea minerals beyond the continental shelf will have to be more specifically defined before United States private enterprise will be induced to engage in these activities. The NPC interim report retorts that adoption of its proposal regarding the definition of the continental shelf would remove any urgency in the foreseeable future to create a new international framework for developing these resources (NPC Interim Report, pp. 16, 17, 19).

It may be true that if the continental shelf is redefined to extend its seaward boundary so as to give each coastal nation permanent, exclusive access to substantial mineral resources of various kinds, relative to its foreseeable

* 1968 ABA Reports, p. 10.

needs in the coming decades, the coastal nations may not be too concerned about the framework for the subsea areas beyond such redefined limits. On the other hand, it may also be true that if the coastal nations are satisfied with the framework for the development of the mineral resources beyond the continental shelves, as redefined in accordance with COMSER's recommendation, they may readily accept that recommendation. As a practical matter, therefore, the question of fixing the outer limits of the continental shelf is inseparable from that of the framework beyond these limits. The two questions will be intertwined in international negotiations.

The NPC position reflects the particular circumstances of offshore petroleum recovery which are not present in the case of other subsea minerals. Malcolm Wilkey, at the meeting of the board of directors of the American Mining Congress, pointed out that the oil industry is not as interested as the mining industry in the international framework for mineral development beyond the continental slope because the oil industry believes most of the oil and gas is to be found in the submarine areas down to, and not beyond, the base of the continental slope. The mining industry, however, "knows that the [hard] mineral resources of the subsoil in the deep ocean [beyond the continental slopes] are likely to be just as rich as on the continental shelf." Indeed, Wilkey is willing to have the United States accept a more narrow continental shelf if that should prove to be the price of obtaining the kind of framework the United States thinks necessary for the subsea areas beyond.

In fact, the NPC itself recognizes that "limited knowledge" makes it impossible to draw a "definitive seaward limit on existence of petroleum deposits" (NPC Interim Report, p. 5). Within two months after this was written, the National Science Foundation announced that the *Glomar Challenger,* on an NSF-sponsored expedition, had drilled a hole penetrating 480 feet into the seabed under 11,753 feet of water in the Sigsbee Knolls area of the Gulf of Mexico and found a show of oil and gas.*

It is not persuasive to argue that a framework for the areas beyond the redefined limits of the continental shelves is not necessary because even if we had such a framework, economic and technological factors would presently inhibit exploration and exploitation of mineral resources in these areas. For it is also true that if technological and economic factors were presently favorable, the existing framework would deter such exploration and exploitation and defeat a basic objective of the Marine Resources and Engineering Development Act.† Just as it takes time—and planning—to prepare the scientific, technological, and economic bases for developing the mineral resources lying deep under water, it also takes time—and planning —to create an international legal-political framework that will be hospitable

---

* *Ocean Industry* 3, no. 10 (October 1968):35.
† Pub. L. 89–454, 80 Stat. 203, 17 June 1966, 33 U.S.C. §§ 1101–08 (1967). COMSER was created by this act.

to such development. Conscious and appropriate lawmaking also will encourage the steps necessary to build the scientific, technological, and economic foundations for the desired activity.

In short, the nations of the world should not underestimate the pace of technological advance in the face of increasing human needs nor the mutually reinforcing relationships that can be formed between scientific, technological, and economic planning on the one hand and conscious and appropriate lawmaking on the other.

Finally, it is important to note that the United Nations is deeply immersed in oceanic matters and that steps have been taken in various of its organs toward redefining the continental shelf and creating a new framework for the exploration and exploitation of the mineral resources of the seabed and subsoil beyond the outer limits of the redefined continental shelf. United Nations activity in this area was given great impetus by the introduction of the Malta Resolution at the twenty-second (1967) session of the United Nations General Assembly.* This activity culminated in the adoption of a resolution at the twenty-third (1968) session of the General Assembly creating a permanent standing committee on peaceful uses of the seabed beyond national jurisdiction (for a summary of the activity leading to this resolution, see International Panel Report, p. VIII–25). Forty-two nations are represented on this committee, their representatives serving six-year staggered terms.

During the debates in the United Nations Political Committee which preceded the adoption of this resolution, there was general consensus that there are other limits to the continental shelves, that these limits must be precisely defined, that a framework must be devised for the development of the mineral resources of the subsea areas beyond these limits, and that all these questions must be considered together.† It should also be recalled that under Article 13 of the Convention on the Continental Shelf, any nation acceding to the convention may request its revision at any time after 10 June 1969. The United Nations General Assembly must then decide what to do about the request.

In light of all these considerations, COMSER recommended that the United States seize the opportunity for leadership which the present situation demands and propose a new international legal-political framework for exploration and exploitation of the mineral resources underlying the deep seas, that is, the high seas beyond the outer limits of the continental shelf as redefined in accordance with COMSER's recommendations. It also outlined the elements of such a new framework (COMSER Report, pp. 147–53). But it took no position on whether the new framework (including the redefined continental shelf) should be embodied in amendments to the Conven-

---

* See U.N. Doc. A/6695 and U.N. Doc. A/6840, Add. 2.
† See UNGA, A/C.1/PV. 1590, 28 October–11 November 1968.

tion on the Continental Shelf, a protocol to the convention, or a separate treaty or treaties. These are questions of tactics best left to those who will negotiate on behalf of the United States.

In determining the elements of its recommended new framework (which will be described below), COMSER considered and rejected a number of alternatives, including proposals to give title to the mineral resources beyond the continental shelf to the United Nations in the name of the international community. Limitations of space preclude a discussion of these alternatives here, but such a discussion may be found in Appendix A of the report of COMSER's International Panel.

An International Registry Authority

According to the COMSER recommendations, all claims to explore or exploit particular mineral resources in particular areas of the deep seas would be registered with an international registry authority. Only a nation, or an association of nations, would be eligible to register such a claim. Every nation adhering to the agreements embodying the new framework would undertake not to engage in, or authorize, exploitation except under a registered claim.

Nations would be free to engage in or authorize preliminary investigations to determine whether it was worthwhile for them to register claims to explore. However, as will become clear later, every nation would have good reason to register exploration claims as quickly as possible.

The membership of the international registry authority and the manner of choosing its governing body would be specified in the agreements embodying the new framework. The authority would find its place in the family of the United Nations, but would be as autonomous as the World Bank. It would be organized on a "multiple principle" of representation, based on the technological capacity of its members as well as on their geographic distribution.

The international registry authority would be required to register claims with respect to specified mineral resources—for example, oil and gas, or all minerals other than oil and gas, or all mineral resources—in specified areas of the deep seas on a "first come, first registered" basis, subject only to the following condition: the nation registering the claim would have to satisfy the authority that the individual, association, corporation, or national organization undertaking the exploration or exploitation is technically and financially competent and willing to perform the task. The entity undertaking the task would not have to be a national of the registering nation—that would be a matter for each nation to decide for itself.

But for the condition mentioned, which is necessary to prevent claim registration from being used to "sit on" the rights derived therefrom, the authority would be given no discretion to deny registration of any claim. Registration of a claim to explore would confer upon the registering nation the exclusive right to engage in or authorize exploration for particular mineral resources in a particular area of the deep seas. The International Panel envisaged that this right would be subject to maximum area limitations and specifications of the period of time within which the resources must be proved (International Panel Report, p. VIII–37). These limitations and requirements would be fixed by the registry authority in accordance with specified standards economically suitable to such exploration. Failure to comply with these conditions, which would also serve to eliminate claim registration for the purpose of "sitting on" rights, would subject the registration to revocation.

Upon proof of discovery, the international registry authority would be required to convert the registered claim to explore into a registered claim to exploit. This claim would confer upon the registering nation the exclusive right to engage in or to authorize exploitation of particular mineral resources in a particular area of the deep seas for a limited number of years. The claim would consequently grant internationally recognized title, or the right to confer such title, to the extracted mineral resources.

The number of years for which the registering nation would have exclusive access to the mineral resources of the area covered by the registered claim, as well as the size of the area, would be fixed by the registry authority. The area would be large enough, and the time long enough, to enable the producer to operate economically and not wastefully and to recover its original investment as well as an adequate return thereon. Each registered claim to exploit would also be subject to the requirements of actual exploitation within a specified period of time and continued exploitation thereafter, unless the registry authority, for good reason, waived these requirements. Failure to comply with the requirements would subject the registration to revocation.

A registering nation would be authorized to transfer any of its registered claims to any other nation which adhered to the agreements embodying the new framework. No unregistered claim would be entitled to any of the benefits derived from registration, and in any conflict between a registered claim and an unregistered claim, the former would prevail. All nations would thus have the incentive to adhere to the recommended agreements and to register claims with the international registry authority.

Upon expiration of the period of registration of a claim to explore or exploit, further exploration or exploitation of the resources covered by the claim would be subject to whatever international legal-political framework was in effect at that time. The nation which registered the expired claim would not acquire, by virtue thereof, a vested right to continue to explore or exploit the particular mineral resources in the area in question. Nor would it even acquire a preference over any other nation with respect to such further exploration or exploitation.

To cover the costs of the international registry authority, every nation would be required to pay to the authority a fee for each exploration claim it registered and an additional fee if and when that claim was converted into a claim to exploit. The authority would be empowered to fix the fees.

An International Fund

Every nation registering a claim to exploit would be required to pay a portion of the value of the production, if any, into an international fund. This money would be used for such purposes as financing marine scientific activity and resources development, particularly food-from-the-sea programs, and aiding developing countries through the World Bank, the United Nations Development Program, and other international development agencies. The international fund would be prohibited from spending the proceeds from these payments for the general purposes of the United Nations. In this way, COMSER sought to insulate its recommended international fund from the debate about whether the United Nations should have an "independent" source of income, particularly for peace-keeping operations.

The registry authority would receive the payments from the registering nations and turn the proceeds over to the international fund, but would have nothing to do with the fund's management. The membership of the international fund, and the manner of choosing its governing body, would be determined by the United Nations General Assembly.

COMSER emphasized that its recommendations with respect to the international fund and its purposes did not constitute just another call upon the rich nations to aid the poor nations. They were intended to suggest a practical way to compensate the common owners of the mineral resources of the deep seas. It is not feasible for all the nations of the world to divide the "economic rent" that should be charged for the limited, exclusive access to deep-sea mineral resources that international registration would provide. The only practical alternative is to use the economic rent for purposes that the international community agrees will promote the common welfare. No better purposes can be suggested than to combat world hunger and malnutrition and to aid developing nations.

COMSER also warned against any optimistic assumption that proceeds from exploitation of the mineral resources of the deep seas will be so huge as to make it unnecessary, in the coming decades, for the rich nations to aid the development of the poor nations in any other way. It pointed out that the international registry authority will defeat its purposes if it fixes the rates of payment to the international fund so high that development of the mineral resources of the deep seas is discouraged.

Enforcement of the Recommended Framework

COMSER's recommendations regarding the enforcement of its proposed framework also mix "national" and "international" elements. Each nation adhering to the agreements embodying the new framework would undertake to enact legislation to implement it and to assure two things in particular: (1) that the business entity on whose behalf a claim is registered complies with the conditions imposed by the international registry authority and reasonably accommodates other uses of the subsea area covered by the registered claim, the superjacent and surface waters, and the air above them, along the lines specified in Articles 3–6 of the Convention on the Continental Shelf; and (2) that the specified fees and payments are submitted to the registry authority. The International Panel further considered the idea that this national legislation should also require the entities on whose behalf claims were registered to collect and submit to the international registry authority such statistics and reports regarding the registered exploration and exploitation activities as the authority might require (International Panel Report, p. VIII–39).

The agreements embodying COMSER's proposed framework would also empower each registering nation to apply its civil and criminal laws to protect the exploration and exploitation activities conducted under its registered claims, including the personnel involved and the necessary installations and other devices, against piracy, theft, violence, and other unlawful interference. The registering nations' failure to discharge these obligations effectively would subject its registered claims to revocation by the international registry authority.

Finally, nothing in the recommended framework would prevent the registering nation from applying any other of its domestic laws, not inconsistent with the recommended framework, to the exploration and exploitation activities under its registered claims, such as laws respecting working conditions, the production and marketing of the extracted minerals, and the taxation of the income from such activities.

COMSER did not recommend that the international registry authority be given initial policy functions. However, because the authority would be empowered to cancel a registered claim if the registering nation failed to discharge its obligations properly, COMSER recommended that the authority be permitted to inspect all stations, installations, equipment, and other devices used in operations under a registered claim and to conduct appropriate hearings. In this way, the authority would have the means to exercise its revocation power fairly and with full knowledge of the facts.

Dispute Settlement

The 1958 Geneva Conventions on the Law of the Sea were accompanied by an "Optional Protocol Concerning the Compulsory Settlement of Disputes." The protocol authorizes any ratifying nation to bring before the International Court of Justice any dispute involving it and any other nation that adheres to the protocol, if the dispute arises out of the interpretation or application of any provision of any of the conventions. The optional protocol excepts disputes subject to the arbitration machinery created by the Convention on Fishing and Conservation of the Living Resources of the High Seas.

The United States has signed but has not ratified the optional protocol. COMSER recommended that the United States ratify the optional protocol. At the same time, however, it proposed special machinery for the settlement of disputes arising out of its recommended framework for deep-sea mineral resource development. It suggested that the international registry authority should initially settle such disputes. At the request of any party to the dispute, however, the authority's initial decision, including a decision to revoke a registered claim, would be subject to review by an independent arbitration agency possessing expertise in resolving the kinds of issues likely to be presented. The International Panel would have the registry authority appoint the members of the arbitration agency (International Panel Report, p. VIII-43).

Creation of an Intermediate Zone

One of the most controversial of COMSER's recommendations calls for the creation of an "intermediate zone," the subsea areas of which would be treated like the subsea areas of the deep seas beyond the redefined continental shelves in all respects, except that only the licensees of the coastal nations would have access thereto. COMSER recognized that the uncertainties surrounding the present definition of the continental shelf may have raised the expectations of some coastal nations to the point where they may refuse to accept its recommended redefinition of the shelf without preferential rights of access to the mineral resources of a reasonable subsea area lying beyond the redefined shelf. It is also realized, to paraphrase the language of the Truman Proclamation of 1945, that self-protection may, for some time to come, compel the coastal nation to keep close watch over activities off its shores which are of the nature necessary for the utilization of the mineral resources lying reasonably beyond the redefined shelf.

Accordingly, COMSER recommended that intermediate zones should be created encompassing the bed and subsoil of the deep seas (that is, the subsea areas beyond the redefined continental shelves) but only to the twenty-five-hundred-meter isobath, or one hundred nautical miles from the base-lines for measuring the breadth of each coastal nation's territorial sea, whichever alternative gives the coastal nation the greater area for the purposes for which the intermediate zones are created. Only the coastal nation or its licensees, which need not be its nationals, would be authorized to explore or exploit the mineral resources of its intermediate zone. Only the coastal nation, therefore, would be authorized to register claims in the intermediate zone with the international registry authority. All other nations and their nationals would be forbidden to enter the intermediate zone for purposes of mineral resource exploration or exploitation without the coastal nation's permission. In this respect, the coastal nation would have the same exclusive authority in its intermediate zone that it would have on its redefined continental shelf. It might decide not to register any claim to explore or exploit mineral resources in the zone, in which case every other nation and business entity would be barred from engaging in such activities in the zone.

At the same time, however, COMSER remained of the view that the mineral resources of the deep seas, including those of the intermediate zones, do not, in fairness or law, belong to the coastal nations, and so all other nations should not be excluded from the benefits of their exploitation. Accordingly, COMSER recommended that in all other respects, including registration with the international registry authority and payments to the international fund, mineral resource development in the intermediate zone should be governed by the framework recommended for the areas of the deep seas beyond the intermediate zones.

The twenty-five-hundred meter/one-hundred-mile alternatives recommended by COMSER as the outer limits of the intermediate zone are about as close together as pairings of world-wide averages of the depths of the bases of the geological continental slopes and the widths of the geological continental shelves and slopes can reasonably be. The boundaries of the intermediate zones, like those of the continental shelves, would be fixed once and for all in terms of geographical coordinates (and by use of a system analogous to that of straight baselines) and would be recorded with the international registry authority. If the same intermediate zone is claimed by two or more nations whose coasts are opposite each other, or by two or more adjacent nations, the boundaries would be determined by applying the median-line principles set forth in Article 6 of the Convention on the Continental Shelf.

Obviously, COMSER proposes a compromise between the position that the continental shelf should be redefined to include the intermediate zone and the position that the intermediate zone should be treated in every respect like the areas of the deep seas beyond it. COMSER did not think that the creation of intermediate zones would raise the dangers it saw in the

NPC proposal to redefine the continental shelf to include the zone. Particularly, it did not think that creation of intermediate zones would encourage claims of exclusive access for purposes other than mineral resource development.

A nation which registers a claim in the intermediate zone (or beyond) would not thereby acquire the "sovereign rights" of a coastal nation over its continental shelf. It would have only the rights accorded it under the new framework. For example, its right of exclusive access would be limited in time; upon expiration of any claim pertaining to its intermediate zone, the coastal nation would be subject to whatever international legal-political framework governed the area at that time. By registering such a claim, it would not acquire a vested right to the mineral resources covered by the claim. Of course, if the framework recommended by COMSER continued to be in effect at the time of expiration, only the licensees of the coastal nation would continue to be eligible to register claims in its intermediate zone.

The coastal nation would pay a portion of the value of production in the intermediate zone into the international fund. The claims of the international community in the intermediate zones are thereby acknowledged. Scientific inquiry concerning the bed of the intermediate zone which would be undertaken there would not require the coastal nation's prior consent. Such consent is of course required in the case of the continental shelf (International Convention, Art. 5 [8]. Finally, the coastal nation would have limited, exclusive access only to the mineral resources of the intermediate zones—and not also to the sedentary living species of the zones.*

The International Panel considered the advisability of an intermediate zone less extensive than the one ultimately recommended. This alternative has a number of advantages. It lends even less credence to the NPC position. It may also help to assure that the claims of coastal nations in the intermediate zones are confined to mineral resources. But acceptance of this alternative may also lessen the attractiveness of the compromise COMSER thought necessary to secure acceptance of its proposed redefinition of the continental shelf. Though I remain open to persuasion on this point, I continue to think the decision of the panel and COMSER was wise.

Other Views on the Framework beyond Limits of the Continental Shelf

The recommendations of COMSER and its International Panel evolved on the basis of studies and discussions carried on inside and outside government. But it may also be said that the panel's tentative conclusions and suggestions also influenced the views of groups in and outside government, and even the official position of the United States.

---

* According to Article 2(4), the "natural resources" to which a coastal nation is given permanent, exclusive access under the Convention on the Continental Shelf include sedentary living species.

The NPC Interim Report. Although the NPC interim report takes the position that a new framework for the exploration and exploitation of the mineral resources beyond the limits of the continental shelf (as defined by it) "will not be required for many years," it suggests the objectives that discussions of such a framework should seek (p. 11). The aim "should not be the creation of an international licensing authority with power to grant or deny mineral concessions but, instead, (a) international agreement on standards of conduct of individual nations [including reciprocal respect for the security of their investments]; (b) international agreement on standards of scope, area, and duration pertaining to development projects, and (c) establishment of procedures for the international recording and publication of their respective claims and activities" (p. 11).

COMSER's recommendations are generally consistent with these NPC views. The international registry authority would not be a licensing authority of the type referred to by NPC. As pointed out above, the registry authority would be empowered to refuse the registration of a claim only to prevent it from being used to sit on the rights derived therefrom. However, the NPC position differs sharply from that of COMSER in not acknowledging the interest of the international community in the subsea areas beyond the continental shelf.

The ABA Resolution. This resolution is also consistent with COMSER's recommendations. Unlike the NPC interim report, this resolution calls for the establishment "as soon as practicable" of a framework governing the areas beyond the continental shelf and one that would not recognize "claims of sovereignty or rights of discretionary control by any nation or group or organization of nations" in these areas.* Like the NPC, the ABA sees as the objective of such a framework "not the creation of a supersovereignty with power to grant or deny mineral concessions, but rather agreement upon norms of conduct designed to minimize conflicts between sovereigns which undertake" exploration and development in the areas beyond the continental shelf. To this end, it urges arrangements that will assure "freedom of exploration by all nations on a nondiscriminatory basis, security of tenure to those engaged in producing the resources in compliance with such rules, encouragement to discover and develop these resources, and optimum use to the benefit of all peoples."

The report of the ABA groups comes even closer to COMSER's recommendations by supporting the payment "in the nature of registration fees, and development fees or royalties" by nations exploiting mineral resources in subsea areas beyond the continental shelf (1968 ABA Reports, p. 13). The ABA groups also would use the proceeds for "international activities on which wide agreement can be reached, such as oceanic research, programs

* See *Proceedings of House of Delegates, American Bar Association,* Philadelphia, Penn., 5–8 August 1968, 54 A.B.A.J. 1017, 1030 (1968).

aimed at improved use of the sea's food resources to alleviate protein malnutrition, and the development of the natural resources of the less developed countries."

ILA American Branch Interim Report. This report made some concessions to COMSER's proposal for the creation of an intermediate zone. While it endorsed the NPC position that coastal nations should join in asserting claims of permanent, exclusive access to the mineral resources underlying the geological continental shelves and slopes and the landward portions of the continental rises, the report also advocated that the coastal nations "may also wish to give serious consideration to provision, in accordance with internal law and constitutional procedures, for allocation of a portion of the revenues derived from part of the area of coastal control to an international fund earmarked for expenditure for generally approved international purposes" (p. xii).

Luke W. Finlay, manager, government relations, Standard Oil Company (New Jersey), objected to this suggestion, saying: "I see no more reason for a nation's allocating a portion of the revenues derived from offshore operations than for allocating a portion of the revenues from onshore operations to international purposes and the very making of the suggestion casts an implied cloud on the title of the coastal states to the mineral resources of their continental margins" (pp xi, xii). Proponents of the COMSER recommendations would agree with Finlay, but would wish that the ILA American Branch had acknowledged the interest of the international community in the subsea areas beyond the continental shelf, as COMSER would define it, and had supported the concept of an "intermediate zone."

For the subsea areas beyond the limits of the continental shelf, however defined, the report rejected the notion of "an international licensing mechanism" with power "to grant or refuse reconnaissance permits, exploration licenses, and production concessions" or "to control or prohibit production, set prices, control repatriation of capital and profits, and fix and collect taxes and royalties" (p. xv, xvi). Again, it should be emphasized that COMSER does not recommend the creation of such an international licensing mechanism. The report came close to approving COMSER's general conception for the framework beyond the continental shelves when it favored establishment of an international commission to draft a convention to accomplish the following objectives:

a. Creation of an international agency with the limited functions of (i) receiving, recording, and publishing notices by sovereign nations of their intent to occupy and explore stated areas of the seabed exclusively for mineral production, notices of actual occupation thereof, notices of discovery, and periodic notices of continuing activity, together with (ii) resolution of conflicts between notices recorded by two or more nations encompassing the same area.

b. Establishment of norms of conduct by sovereign nations with respect to the recording of the notices proposed in the preceding paragraph, and in the occupation of the seabed and exploration and production of minerals therefrom. The drafting commission could appropriately recommend for inclusion in the resulting convention, among other things, standards (or a mechanism to establish standards) relating to permissible areas for inclusion in exploration and production phases, periods of exclusive rights of occupancy, requirements of diligence as related to tenure, conservation, avoidance of pollution, accommodation with competing uses of the marine environment, etc. The instructions to the negotiating commission should stipulate that the resulting convention shall contemplate that the actual production and marketing of minerals discovered shall be controlled by the laws of the recording nation, and that that nation shall be held accountable for the conduct of those operating under its flag in the exploration and exploitation of minerals.

c. Establishment of (i) reasonable payments to be made, preferably to the World Bank, by the nation which undertakes mineral development, in areas seaward of coastal mineral jurisdiction, in the nature of development fees or royalties and (ii) the purposes to which such revenues, when received, shall be applied. These purposes should be restricted to international activities on which wide agreement can be reached, such as oceanic research, programs aimed at improved use of the sea's food resources to alleviate protein malnutrition, and the development of the natural resources of the less developed countries. [Pp. xv, xvi]

Resolution of the American Mining Congress. Although the resolution adopted by the American Mining Congress Board of Directors on 6 October 1968 did not deal with the subject, Malcolm Wilkey, speaking for the board, opposed the idea of having "some sort of a royalty paid on the product lifted from either the continental shelf or the deep ocean subsoil." However, he would not object to "paying appropriate fees to finance the administration of an international mining claim registry office, but the fees should be no larger than sufficient for this purpose."

The Mining Panel of the National Security Industrial Association's Ocean Science and Technology Advisory Committee (OSTAC) has argued that there is no reason for revenues from deep-ocean mining to be treated differently from revenues from ocean fishing because ocean fish are as much the common property of mankind as are ocean minerals.* But at present, no nation has the right of exclusive access to any high seas fishery beyond the widely recognized twelve-mile exclusive fisheries zone. Registration with the proposed international registry authority would give the registering nation limited rights of exclusive access to ocean minerals in return for which it is equitable to require it to pay an "economic rent" to the common owners of the minerals.

Indeed, the absence of rights of exclusive access to the fisheries of the high seas accounts for many of the serious difficulties facing the nations engaged in these fisheries. COMSER's recommendation for the institution—on

* Comments on COMSER Report by Mining Panel, Ocean Science and Technology Advisory Committee, National Security Industrial Association (hereafter OSTAC Comments), April 1969.

a trial basis—of national catch quotas for the North Atlantic cod and haddock fisheries is an attempt to solve these difficulties by creating rights of exclusive access (COMSER Report, pp. 105–10). The International Panel suggested that once such a quota system is instituted, a portion of the net economic gain realized from the right of exclusive access thereby accorded might be paid into the international fund set up in connection with the exploitation of ocean minerals (International Panel Report, p. VIII–64).

Unlike Wilkey, however, the OSTAC Mining Panel also opposes creation of the recommended international registry authority. It fears that the "lack of knowledge of the nature, location and value of deep ocean minerals (particularly nodules) could lead only to a wild race and scramble to register claims to very extensive areas by competing nations and nationals with no sound or established basis by the registering authority for making and enforcing decisions that are likely to be either challenged or ignored." Yet the Mining Panel goes on to suggest "an international agency for recording, as a matter of general information, intentions to explore particular areas which would be subsequently exploited under rules and standards including safety and control of pollution, etc., that would be established by each country for its nationals whose operations would be undertaken and protected under the present concepts of the 'freedom of the seas.'" In the Mining Panel's opinion, there are "adequate, existing international courts and mechanisms for dealing with disputes in an orderly and peaceful way" (OSTAC Comments).

If I understand the Mining Panel correctly, it maintains either that it is not necessary to grant limited rights of exclusive access in order to encourage exploration for and exploitation of manganese nodules, or that existing international law grants such rights to the first discoverer. Neither of these views, in my opinion, can be sustained. International fishing, to which the Mining Panel looks as a model of the application of the concept of the freedom of the seas, is plagued with difficulties that arise from the lack of limited rights of exclusive access.

Nineteenth Report of the Commission to Study the Organization of Peace. The most recent report of the Commission to Study the Organization of Peace (CSOP) would go further than COMSER toward the creation of an international authority to regulate exploration and development of undersea mineral resources.* It recommends, first, that the United Nations General Assembly promptly adopt a declaration of principles to govern such exploration and development (p. 21). The principles it suggests are generally in line with those proposed by COMSER. The declaration would call for the definition "with all possible speed" of a precise boundary for the area be-

---

* *The United Nations and the Bed of the Sea.* Nineteenth Report of the Commission to Study the Organization of Peace (hereafter CSOP Report), Washington, D.C., March 1969,

yond the "limits of national jurisdiction," and require that this area "should be as large as possible so as to preserve the largest amount of resources for the benefit of mankind and to diminish the possible area of controversy" (p. 21). It would also ask for the creation of "an appropriate international regime, established by the United Nations" for this area which should "allocate leases on the basis of competitive criteria" and "take into account the economic interests of the developing States" and make "arrangements for dedicating a reasonable portion of the value of such resources to international community purposes, including the economic, social, scientific and technological progress of the developing countries" (p. 22).

The CSOP report goes on to recommend, but not for purposes of inclusion in the United Nations declaration of principles, that the continental shelf be redefined as proposed by COMSER (p. 24). It would have no "intermediate zone," but would give jurisdiction over the region beyond the two-hundred-meter/fifty-mile line to a United Nations authority (p. 25). Unlike COMSER, too, the CSOP report recommends that "the United Nations appeal to all States to refrain from granting leases to mineral resources beyond the 200-meter depth limit until the revision of the Convention on the Continental Shelf comes into force" and that the United States "respond favorably to such an appeal and . . . use its good offices to obtain favorable responses from other States" (p. 25).

The international authority recommended by CSOP is also different from that proposed by COMSER. It would be established by the United Nations General Assembly; its executive head would be named by the United Nations secretary-general; and it would be directed by a council of states selected by the General Assembly (pp. 27–28). The authority would be empowered to "manage the leasing of the sea-bed beyond the recognized national jurisdiction of all States" through, *inter alia,* a system of competitive lease sales (pp. 27–28).* The authority would also "ensure the use of the sea-bed for peaceful purposes only."

---

* The CSOP proposal is very much like that embodied in the suggested *Treaty Governing the Exploration and Use of the Ocean Bed* prepared by the United Nations Committee of the World Peace through Law Center (Pamphlet Series No. 10, undated). The treaty, however, does not require its ocean agency to license exploration and exploitation solely on the basis of competitive bidding. It also authorizes the agency to disregard the highest bid requirement in order to grant a license to a developing nation so as to improve its capability, or that of its nationals, to explore or exploit ocean-bed resources. The CSOP Report, on p. 15, criticizes this aspect of the treaty as follows: "It has been stated by some that the technologically less developed States might share directly in the wealth of the sea-bed by receiving preferential treatment in the allocation of exclusive rights. That is, they might receive such rights on the basis of need rather than on the basis of full payment of royalties or bids. Such a system, however, would work neither to the benefit of the world community nor to the benefit of the recipients of the rights. Royalties paid to the world community would be diminished. And the less developed States would be encouraged to make

The International Panel—and COMSER—considered and rejected the CSOP proposal that a United Nations agency be empowered to auction to the highest bidder exclusive rights to explore and develop the mineral resources of particular deep-sea areas. Two principal considerations led to this rejection. The International Panel feared that nations would bid for political, technological, prestige, security, or economic warfare reasons, thereby increasing international rivalry and conflict. Secondly, because some bidders might be nations and others private entrepreneurs, economic rationality might not necessarily be reflected in their competitive bidding.

United States Position on Framework beyond Limits of Continental Shelf. The position of the United States with respect to the international framework for the exploration and exploitation of subsea minerals has undergone significant and desirable change during the life of COMSER (see International Panel Report, pp. VIII-28, 29, 31) and now points to the COMSER recommendations. The United States has proposed that the United Nations General Assembly adopt a resolution commending to nations "for their guidance" the following principles concerning subsea mineral exploration and exploitation: *

1. The outer limits of the "continental shelf," for purposes of the Convention on the Continental Shelf, should be defined precisely "as soon as practicable." The Subcommittee on International Organizations and Movements of the House Foreign Affairs Committee has called for "prompt action" to this same end.†

2. Exploitation of subsea mineral resources "that occurs prior to establishment of the boundary $\pm \beta$ of the continental shelf] shall be understood not to prejudice its location, regardless of whether the coastal [nation] considers the exploitation to have occurred on its 'continental shelf.' "

3. No nation may claim or exercise sovereignty or sovereign rights over any part of the seabed or subsoil beyond the redefined limits of the continental shelf.

4. The interest of the international community in the development of mineral resources beyond the redefined shelf should be recognized through

---

inefficient use of their scarce technical skills and capital. Such investments would have little payoff to the growth of their economies."

In the light of this criticism, it is not clear what the CSOP report had in mind when it recommended that its licensing agency should "take into account the economic interests of the developing States." CSOP Report, p. 22.

* "Draft Resolution Containing Statement of Principles Concerning the Deep Ocean Floor," U.S. Mission to the United Nations, Press Release U.S.U.N., 107 (68), 28 June 1968. The draft resolution is set forth verbatim in International Panel Report, pp. VIII-30–31.

† *The Oceans: A Challenging New Frontier.* H. Rep. No. 1957, 90th Cong., 2d Sess. 3R (1968). The Subcommittee on International Organizations and Movements was informed of the work of the UN Ad Hoc Committee and the positions taken by the United States in that forum.

the "dedication as feasible and practicable of a portion of the value" of these resources to "international community purposes."

5. A new framework should be established "as soon as practicable" for the exploration and exploitation of the mineral resources beyond the redefined continental shelf which will be conducive to the making of the investments necessary to conduct these activities.

A Proposed Course of Interim Action. COMSER recognized that it will take years to negotiate new arrangements for the development of subsea mineral resources and that it is important for the nations of the world to come to the earliest possible agreement on the policies by which they will be guided in the interim. Accordingly, COMSER supported the principles that the United States proposed for adoption by the United Nations General Assembly, but recommended an important addition (COMSER Report, pp. 155–56).

The United States proposals quite properly seek to reserve the subsea areas beyond the two-hundred-meter isobath for future international decision. But they do not indicate what the coastal nations should reasonably consider to be the limits of their continental shelf until such time as the shelf's boundary is fixed by new international agreement. Nor do they say how the United States will define the limits of its continental shelves until that time.

Subsequent to the publication of the COMSER report, the State Department, still speaking for the Johnson administration, indicated that "it had been agreed within the Executive Branch that at the present time the United States should take no position as to the outer boundary of the continental shelf or the regime for the area beyond." * Nor has the Nixon administration as yet taken any position on the matter.

Until new arrangements are agreed upon, it is possible that coastal nations may claim wider continental shelves than would be consistent with United States interests and thereby influence the ultimate location of the shelf's boundary. COMSER recommended, therefore, that the United States propose the principle that in the interim no nation should claim or exercise sovereignty or sovereign rights over any part of the seabed or subsoil beyond the two-hundred-meter isobath (COMSER Report, p. 156).

This was not intended to imply that there should be a moratorium on exploration and development beyond this isobath until the new international arrangements are negotiated. On the contrary, COMSER rejected the suggestion of a moratorium which has been made by a number of nations in the course of the United Nations discussions. COMSER recommended that the United States continue to authorize exploration and exploitation of the

---

* Department of State, Transcript of Press and Radio News Briefing, 14 January 1969. The statement read to the news media was prepared by an Interdepartmental Committee on International Policy in Marine Environment.

mineral resources of the seabed and subsoil underlying the high seas beyond the two-hundred-meter isobath, on condition that such authorization explicitly state that any such exploration or exploitation shall be subject to the new international framework eventually established (COMSER Report, p. 156).

No other policy will achieve the objective of the Marine Resources and Engineering Development Act to preserve "the role of the United States as a leader in marine science and resource development (COMSER Report, p. 156). Nor will it benefit mankind to postpone subsea mineral resource development for an indefinite period of time.

Another alternative is to have any new, international agreement recognize "grandfather rights," that is, to permit those business entities which, in the interim, develop mineral resources beyond the two-hundred-meter isobath to continue to do so on the same terms and conditions that induced them to undertake these activities. The commission rejected this alternative because it would create incentives to postpone international agreement and prejudice the ultimate location of the shelf boundary. Apparently this alternative also did not appeal to the United Nations General Assembly.

The commission recognized that if exploration and exploitation proceeded beyond the two-hundred-meter isobath, in accordance with its recommendations, private industry would run the risk of ultimately finding that its operation was located in an area beyond the redefined continental shelf and, perhaps, governed by a less favorable framework than now applies on the outer continental shelf. Yet United States objectives in the oceans would be thwarted if, to avoid the risk described, American private enterprise declined to proceed with exploration and exploitation beyond the two-hundred-meter isobath. To resolve this dilemma, the commission recommended that Congress enact legislation to compensate private enterprise for loss of investment or expenses occasioned by any new international framework that redefines the continental shelf so as to put the area in which it is engaged in mineral resource development beyond the shelf's outer limits.

Wilkey has argued, again at the aforementioned American Mining Congress meeting, that such compensation would be insufficient to encourage mining beyond the two-hundred-meter isobath. The mining industry "necessarily had to have a far greater return from any successful venture than just compensation for the money expended, because there were always unsuccessful ventures that had to be paid for by the small percentage successful." The OSTAC Mining Panel has voiced the same criticism as Wilkey.

I sympathize with Wilkey's complaint. But it is not likely that United States private enterprise will run serious risk by venturing beyond the two-hundred-meter isobath if the international framework ultimately agreed upon is patterned upon COMSER's recommendations. (The reasons for this conclusion will become clear in the light of the discussion in the next section of the paper.) Nor is it likely that the United States will agree to any other new framework that will greatly defeat the expectations of private industry, which was encouraged by the United States to proceed with mineral development in deeper waters. These considerations lead me to conclude that Congress might reasonably go beyond COMSER's recommendations and enact legislation to compensate private enterprise for loss of expected profits, as well as investment and expenses, occasioned by the framework ultimately negotiated. Such legislation would afford positive encouragement to private industry at little or no cost to the government.

The State Department, still speaking for the Johnson administration, interpreted its interim position as requiring that "in connection with any arrangements agreed to [for a new international framework], protection must be given to the integrity of investments in areas which lie seaward of the boundary established for national jurisdiction." * It is not clear from this statement, however, whether the State Department contemplates the type of compensation recommended by COMSER or the system of "grandfather rights" rejected by COMSER. As yet, the Nixon administration has taken no position on this matter.

In any case, it is important that private entrepreneurs should know when they must seek the government's permission to engage in exploration and exploitation of subsea minerals and that the United States should have adequate control over the evolving situation until a new international framework is adopted. At present, the Outer Continental Shelf Lands Act is as uncertain on this matter as the existing definition of the continental shelf.† To eliminate this uncertainty, COMSER recommended that the act be amended to require permission from the secretary of the interior to engage in mineral resource exploration or exploitation in any subsea area beyond the two-hundred-meter isobath upon such terms and conditions as the secretary may deem appropriate. It urged that the amendment should make clear that this requirement was not intended to constitute a United States claim to exercise sovereignty or sovereign rights over any subsea area beyond the two-hundred-meter isobath (COMSER Report, pp. 156–57).

UNITED STATES REGISTRATION OF BUSINESS ENTITY CLAIMS

As stated above, nations adhering to the new agreements embodying the framework recommended by COMSER would be obliged to enact implementing domestic legislation. Apart from this obligation, the new agreements would leave it to the nations registering claims with the international registry authority to determine their relations with the business entities on whose behalf the claims would be registered. Nevertheless, COMSER had to

---

* Department of State, Transcript of Press and Radio News Briefing, 14 January 1969.
† 67 Stat. 462, 43 U.S.C. §§1331–43 (1964).

consider these domestic relations in discharging its statutory mandate (COMSER Report, pp. 153–55). It concluded that new legislation would be necessary, in addition to the amendment of the Outer Continental Shelf Lands Act mentioned above, to fix these relations and implement the recommended framework.

COMSER recommended that the new legislation fixing these relations be based on the policies the United States presently follows in leasing mineral resources on its outer continental shelf, with some important modifications.

Policies Applicable to All Registered Claims

Business entities, domestic or foreign, which sought to have the United States register claims on their behalf with the international registry authority would apply to the Department of the Interior, which COMSER recommended as the agency for this purpose. The secretary of the interior would have discretion to determine whether to register the claim as applied for. In exercising his discretion, he would use criteria consistent with those the international registry authority itself would use and would also be guided by the State Department's judgment as to the foreign policy implications, if any, of registering the particular claim.

The business entity on whose behalf a claim to explore or exploit was registered would pay to the United States the specified fees that the United States would be obligated to pay to the international registry authority. The business entity on whose behalf a claim to exploit was registered would also pay to the United States the portion of the value of the extracted minerals which the United States would be obligated to pay to the international registry authority for the international fund. This payment would take the place of both the fixed annual rent per acre or square mile and the royalty on the value of production which must now be paid under mineral leases on the outer continental shelf. In this way it was intended to recognize the rights of the international community to the mineral resources of the deep seas without unduly burdening United States private enterprise.

It is conceivable that these payments for the international fund might be greater than the combined rents and royalties paid for the same mineral value extracted from the outer continental shelf. Because the United States representative on the international registry authority would participate in fixing these payments and the authority itself would be eager to encourage mineral resource development in the deep seas, the payments are not likely to be greater than such combined rents and royalties; it is conceivable they might even be less.

Additional Policies Concerning Claims in the Intermediate Zone

The recommended framework gives the United States, and every other coastal nation, valuable rights in the intermediate zone because only coastal nations may register claims there. It is entirely equitable, therefore, that the United States should sell this right of exclusive access to the highest bidder. It is not essential, however, that the bidding should be in terms of fixed amounts of cash to be paid before exploitation begins, as is now the case on the outer continental shelf. It might be in terms of a percentage of profits or net return.

Furthermore, so long as a single nation has exclusive access to certain mineral resources, a bidding procedure tends to allocate the resources to the most efficient producer and, at the same time, to induce competing applicants to increase their bids to the point where the return on capital is not above the competitive level. For the same reason, differences between the rents and royalties paid on the outer continental shelf and the payments to the international fund from the intermediate zone are likely to be reflected in different levels of competitive bidding for mineral deposits of estimated equal value.

However, as certain minerals are sought in deeper waters, the costs of exploration and evaluation may become so high as to deter these activities. To achieve the objectives of the Marine Resources and Engineering Development Act, COMSER recommended that the secretary of the interior should be authorized to waive competitive bidding whenever he determined that this step was necessary to encourage exploration and exploitation of hard minerals on the outer continental shelf (COMSER Report, pp. 136–37). It was not thought that petroleum exploitation on the outer continental shelf needed this encouragement, but it may be necessary for deep-sea operations. When competitive bidding is waived, the business entity which discovers the deposits will be given the right to exploit them, subject to terms of compensation stipulated in advance.

COMSER recommended that the secretary of the interior should be given similar discretion in the intermediate zone. Unfortunately, its report does not make it clear that this discretion was intended to be granted for the development of oil and gas, as well as hard minerals, in the deep seas.

In every case, a claim to explore would be registered on behalf of the first responsible qualified business entity that applied. If competitive bidding followed discovery, the registered claim to explore would be converted into a claim to exploit on behalf of the responsible qualified business entity that bid the highest cash bonus, or percentage of profits, therefor. That entity might not be the entity on whose behalf the claim to explore was registered. But the registering nation would be authorized to substitute any responsible qualified business entity for the entity on whose behalf the claim to explore was originally registered. If competitive bidding were waived by the secretary of the interior, the registered claim to explore, following discovery, would be converted into a claim to exploit for the entity on whose behalf the claim to explore was originally registered.

Additional Policies Concerning Claims beyond the Intermediate Zone

For claims in deep-sea areas beyond the intermediate zone, COMSER recommended that the United States not use competitive bidding, but adopt the same policy of "first come, first registered" that would guide the international registry authority. Accordingly, the United States would register a claim to explore on behalf of the first responsible qualified business entity that applied. Upon discovery, the claim to explore would be converted into a claim to exploit on behalf of that same entity.

COMSER here sought to avoid a "flag-of-convenience" problem. No such problem can arise in the intermediate zone because the coastal nation would have exclusive access to the zone's mineral resources. But this problem could arise if the United States used competitive bidding in the deep seas beyond the intermediate zone. All nations have equal access to these areas. Explorers and exploiters would then seek to have claims on their behalf registered by nations which charged less or nothing for the privilege. The United States could forbid its nationals, or foreign business entities controlled by its nationals, to have any other nation register claims on their behalf. But COMSER feared that this would establish an undesirable precedent and preferred to await experience with its recommended alternative of no competitive bidding.

Speaking for the American Mining Congress, Wilkey questioned the wisdom of the COMSER recommendation that only nations or associations of nations should be authorized to register claims with the international registry authority. "This introduces an unnecessary link in the chain," Wilkey said, "and might very well be susceptible to political abuse, or be suspected of being politically abused." He thought United States business entities should be permitted to deal directly with any international authority.

Unlike the United States fishing industry, the United States petroleum and mining industries are accustomed to dealing with other nations directly, and not through their own government. Wilkey's position, therefore, is understandable. But under COMSER's recommendations, the United States government would assume important powers and obligations with respect to the development activities of its nationals or other licensees in the subsea areas beyond the continental shelf. For example, the United States would be required to assure that these activities did not unreasonably interfere with other uses of the oceans and that the business entities on whose behalf claims were registered complied with the conditions of registration. If the United States, and not the international registry authority, is to be responsible for these policing functions and if claim registration is not to have adverse foreign-policy consequences, the United States must be given discretion to determine the business entities on whose behalf claims will be registered. Giving it the sole power to register claims will assure the exercise of this discretion. Furthermore, under COMSER's recommendations, only licensees of the United States would have access to the intermediate zone and would-be licensees in this zone, of necessity, would have to deal with the United States.

COMSER's recommendation that the leasing policies on the outer continental shelf, with the modifications indicated, be followed in the areas beyond the redefined continental shelf has given rise to some questions which need to be clarified. Under existing United States law, oil produced on the United States continental shelves is not subject to the system of oil import quotas; but oil produced by United States companies on the continental shelves of other nations is subject to the import quota system. Furthermore, the production of oil from the submerged lands of the states of the United States is usually prorationed. Production of oil on the outer continental shelf is subject to United States regulation.

In making its recommendations, COMSER did not mean to suggest that oil produced in the subsea areas beyond the redefined continental shelves should not be subject to the oil import quota system or to other regulation. COMSER pointed out that the production of oil in these subsea areas would be encouraged if exempted from the quota and other regulatory systems (COMSER Report, p. 126). But it saw that such exemption would raise complex economic, political, and military questions which it was not equipped to analyze. For this reason, it contented itself with a call for "a thorough new assessment of the adequacy of the Nation's offshore and land oil reserves" as the foundation "for shaping the incentives to explore and develop subsea oil reserves and for establishing an orderly, rational leasing policy pacing development at a rate that is in the public interest" (p. 126). Nothing in the sections of the COMSER report dealing with the international framework—and nothing in the report of the International Panel— was intended to suggest policies to be followed on these matters before the new assessment is made.

CONCLUSION

COMSER's recommended international legal-political framework for the exploration and development of the mineral resources of the bed of the high seas and their subsoil is intended to meet the needs of the immediate future, not to suffice for all time. It does not foreclose the following possibilities as: (1) a United Nations agency might, in time, be given title to the mineral resources of the deep seas; (2) each coastal nation might be given permanent, exclusive access to the mineral resources of a submarine area off its coast which is larger than the area of the continental shelf as redefined in accordance with COMSER's recommendation; (3) any nation might be given permanent, exclusive access to the mineral resources of the areas of the deep seas which it is the first to discover and exploit; or (4) some other alternative not now envisaged might be adopted.

COMSER's recommendations probably foreclose two future possibilities which are contrary both to American objectives in the oceans and to the interests of the international community as expressed in the United Nations: that the coastal nations might divide the mineral resources of the deep seas among themselves, and that a nation might claim territorial sovereignty over areas of the deep seas, preventing any use of the areas without its prior permission.

For the immediate future, COMSER concluded, the recommended framework and accompanying national policies would help to achieve American objectives in the oceans. They would encourage the scientific and technological efforts and the other major capital investments needed for exploration and exploitation of the mineral resources of the bed of the deep seas and their subsoil by creating machinery for the international recognition of claims to exclusive access to such mineral resources in large enough areas for long enough periods of time to make operations profitable. They would give American private enterprise fair opportunities to share in such exploration and development.

The recommended framework would minimize international conflict. While the "first come, first registered" principle governing the international registry authority might stimulate a race among nations to register claims to the mineral resources of the deep seas, the recommended framework would greatly temper the nature of this contest. Most important of all, it would impose time limits on registered claims, upon the expiration of which further exploration or exploitation of the mineral resources in the area of the expired claim would be subject to whatever legal-political framework might then be in effect.

Through the recommended international fund, the poor and developing nations of the world would share the benefits of exploration and exploitation of the mineral resources of the deep seas. And finally, the recommended framework would be subject to change in the light of experience with mineral resources exploration and exploitation in the deep seas—subject only to the time-limited claims already registered.

On the whole, the recommended framework is a necessary, first step to assure, in former President Johnson's words, that the "wealth of the ocean floor [is] freed for the benefit of all people."

REFERENCES

Alexander, L. M. (ed.)
1969   *The Law of the Sea: Offshore Boundaries and Zones.* Columbus: Ohio
       State University Press.
Bernfeld, Seymour S.
1967   "Developing the Resources of the Sea: Security of Investment." *International Lawyer* 2, no. 1, pp. 67–76.
Commission on Marine Science, Engineering and Resources (COMSER)
1969   *Our Nation and the Sea: A Plan for National Action.* Washington,
       D.C.: U.S. Government Printing Office.
Goldie, L. F. E.
1968   "The Contents of Davy Jones's Locker." *Rutgers Law Review* 22,
       no. 1, pp. 1–66.
Henkin, Louis
1968   *Law for the Sea's Mineral Resources.* Prepared for the National Council on Marine Resources and Engineering Development. ISHA Monograph No. 1. New York: Institute for the Study of Science in Human Affairs, Columbia University.
International Panel of the Commission on Marine Science, Engineering and Resources
1969   *Report of the International Panel of the Commission on Marine Science, Engineering and Resources.* Washington, D.C.: U.S. Government Printing Office.
Oda, Shigeru
1967   "The Geneva Convention: Some Suggestions to Their Revisions." Paper given at the National Institute on Marine Resources, American Bar Association, at Long Beach, Calif., 7–10 June.
Oxman, Bernard H.
1968   "The Preparation of Article 1 of the Convention on the Continental Shelf." Unpublished memo.

New Machinery for
Policy Planning in Marine Sciences

EDWARD WENK, JR.

IN presenting an overview on new developments in blending oceanography with public policy, from the perspective of the Capitol, I have divided the material into two sets of considerations: the first concerns the decision-making process; the second concerns substance, or the issues now devolving in oceanography. The mixture of oceanography and public policy has led those of us responsible for implementing the landmark legislation of 1966 to substitute a new phrase for oceanography, as reflected in the titles of the reports transmitted recently by the president to the Congress. The field is being termed "marine science affairs," which is shorthand for marine science and technology related to public affairs. This designation reflects translation of the mandate of the Marine Resources and Engineering Development Act of 1966 into action terms whereby we extend the previous emphasis on understanding the oceans to a higher state of policy considerations, that of trying to make more effective use of the oceans and their resources.

Some of our more visionary oceanographers have proposed this evolution for a long time. What is really new is a statutory dedication of a national policy to that purpose. Those of you who have read this act of 1966 find it brief; but it is also one of the clearest pieces of legislation I have read, understandable even to the nonlawyer. It states our objectives and assigns responsibility for implementation of a complex set of tasks that have never been woven together before in any single piece of legislation. The office selected for such implementation is that of the president of the United States. The act reflects the fact that it is the federal government which has the key

Edward Wenk, Jr., is professor of engineering and public affairs at the University of Washington. At the time this article was prepared in 1968, he was serving as executive secretary of the National Council on Marine Resources and Engineering Development.

responsibility to reap benefits of the sea in meeting national needs; and since that responsibility is spread among eleven departments and agencies, there was only one man on whom the Congress could fix that otherwise diffuse responsibility—that is, the president.

I shall try to summarize the president's responsibilities and the pieces of apparatus that were given him to help carry out the job, and in particular the policy-planning and coordinating council with which I am associated. I mentioned trying to approach this question in two complementary dimensions. If one tries to find an analogy, for purposes of explanation, it may lie in a piece of audio equipment. Any of you who has taken an interest in high fidelity may find your fascination split between building and wiring together your own equipment on the one hand and listening on the other. What I am going to try to do is give you a wiring diagram of this apparatus in Washington, but also portray some of the music that goes through the circuits. You have to decide for yourself on the degree of fidelity.

The starting point for diagraming this activity is the office of the president and the National Council on Marine Resources and Engineering Development. This policy-planning group to assist and advise the president is chaired by the vice-president and includes the heads of all of the major agencies that are involved in oceanography: the secretaries of state; commerce; interior; health, education and welfare; transportation; and the navy; and the heads of the National Science Foundation and of the Atomic Energy Commission. Several others with minor roles have been included as observers: the secretary of the Smithsonian Institution, the administrator of the National Aeronautics and Space Administration, and the administrator of the Agency for International Development. It is significant that the muscle for carrying out marine-related programs lies in these operating agencies. You might consider them as a cluster of boosters on some sort of a rocket. And with that analogy, the council corresponds to the rocket guidance system.

To complete the wiring diagram, it is necessary to include the other players in the game. First, there are other parties in the executive branch who share in advising and assisting the president. Thus, included as additional observers are the director of the Bureau of the Budget, director of the Office of Science and Technology, and chairman of the Council of Economic Advisors.

The legislation provides for an independent council staff under its executive secretary to carry out whatever tasks are assigned. Following a well-known principle of any management organization, one seeks inputs from specialized committees; and so, after the first year, these were duly created. Interestingly enough, though composed of representatives from the participating agencies that report directly to the president, these committees report back through the council, serving as a "back door"—a somewhat unorthodox arrangement in government, but effective in picking out projects rejected by individual operating agencies but needed to serve government-wide goals.

This array of executive branch entities does not complete the story. We immediately must add a very significant counterpart, the Congress, in which this initiative to strengthen oceanography in fact began. There are thirty-one subcommittees of the Congress that deal with the oceans. I do not know whether we should be impressed or overwhelmed with that fact, but it does give some idea of the past fragmentation of responsibilities and the complexity of relationships between the executive and legislative branches. One recognized fact of political life incidentally is that while agency heads report to the president and presumably have but that one loyalty, private arrangements are often made with congressional committee chairmen without presidential clearance. Running the government from the White House becomes something of a myth, because these cross-connections multiply all the time.

These internal elements of government do not complete the wiring diagram because it is the people who are being served. Thus, it is necessary to include the institutional groups outside of government who are involved, the universities and academia, industry, and state government. Moreover, there is nothing homogeneous about an academic group; and diversity is also found in industry. Not only do individual concerns compete with each other, but there are characteristically different classes of industry: the aerospace people, who are building research submarines or instruments, have quite different interests and viewpoints than the people who produce oil and gas off the continental shelf. Moreover, every state operates independently and has a variety of well-developed relationships with the federal government.

The story is still not finished. We must consider another set of players in this game—the 111 other nations who front on the sea. The continents cover 29 percent of the planet, and the unique feature of the maritime remainder is that this territory does not belong to any nation. We have a long legal history of freedom of the seas, and we have some history of collaboration among scientists of different nations. And to digress for a moment, it is now a matter of deliberate United States policy to do everything possible to make the exploration of the oceans a global, multinational affair, and this necessarily brings in a number of countries.

Then we must consider another set of interested intergovernmental organizations: the United Nations, the United Nations Educational, Scientific, and Cultural Organization (UNESCO), the Food and Agricultural Organization (FAO), and the World Meteorological Organization (WMO). I was surprised to find that these bodies have difficulty communicating with each other, just like the federal agencies. As a matter of fact, the rivalry between these international organizations makes trying to get anything done internationally more difficult than seeking interagency cooperation in a single government.

In order to understand this wiring diagram, we must add still other components. If we are dealing with, say, agriculture, the farmers, the farm bloc, the food processors, the fertilizer manufacturers, state interests through the experiment stations or whatever can all be identified as connected interest groups. All of these groups cultivate their own connections with committees in the Congress, as well as with the department looking after their concerns.

We do not yet have citizens living under the oceans, and the fish do not vote. The consequence is that with this complexity of institutions involved in marine science affairs, there is not a strong lobby. I use the word "lobby" in a purely declarative sense, not editorial, in the belief that a pluralistic society operates as a series of pressure groups which produce equilibrium and balance some way or other in national policies. All these forces, incidentally, impinge on Washington.

The question concerning stimulus for marine science affairs that motivated the new legislation, therefore, is how you get an enterprise going in the absence of a lobby and the absence of a crisis. We have no wet Sputnik, no counterpart to the Soviet threat to use space technology as an instrument of national policy. We have no special interest group electrifying Washington with signals that this field ought to be moving faster. Now you begin to see the problem presented by the legislation, which says, in effect: "Let's get on with the job: let's accelerate our studies of the sea and use that knowledge in every way possible for national defense, for the development of oil, gas, and minerals off the continental shelf, for the development of fisheries and other living resources, and for the enhanced use of our coastal zones, where the people and the oceans meet." The coastal zones in fact turn out to be a vastly neglected but important arena for policy and for research, an overlooked interface as significant as that between air and sea which scientists already recognize as crucial to predicting weather and, perhaps someday, climate. In this spectrum of uses, we cannot forget to look at the sea as a scientific laboratory and as a medium for international cooperation.

With all of these different, rich, exciting, challenging objectives, how then, in the political process, do you get this enterprise to move? No one solved that enigma before they wrote the legislation; I am reasonably sure they did not construct the wiring diagram presented here. In fact, the scientific community approached this question of benefits as a means simply to expand research. But, the fact of the matter is that the law gives the president the power and discretion to utilize the sea more effectively; and until some revision of that legislation developed out of the recommendations of the Stratton Commission, the president had the sole responsibility for trying to get this program moving.

Given the complexity of different participants, the variety of different and sometimes conflicting goals, and the absence of any crisis or pressure from a lobby, the leadership now vested in the president, vice-president, and council secretariat has attempted to move the enterprise by pure logic. Washington is not known for having a very rich history of accomplishments on the basis of pure logic. Therefore, on statistical terms, the prognosis of this particular enterprise is uncertain. But what else have we? So let me then tell you what the council has tried to do since August 1966, when it was created.

In the first instance, the council is not a paper organization. It has met eleven times, with the vice-president in the chair, and it has adopted an activist course. Any of you who are familiar with interagency bodies know about groups that simply meet together to find some common agreement by what is often referred to as "the least common denominator." Many of these bodies never become very innovative or even effective. It turns out, however, that given the charge from the Congress in addition to the energy and personality of Vice-President Humphrey, the council immediately adopted a program of trying to recommend goals and milestones by which this field could be given a list of targets and suggestions of specific actions on how to reach them.

The technique adopted is one of identifying new initiatives. I realize this involves a semantic difficulty, because by and large there is not anything really new in the world. But if ideas went unused and were suddenly picked up again, maybe it is fair to say the initiative is new. In any event, the style of operations has been to try to identify the priority goals that deserve attention for social or economic reasons and that are feasible in terms of today's science and technology. Even that process of policy planning does not guarantee achievement because of the next steps of utilizing the institutions necessary for a graceful and effective transition of science into political action. Let me recall how often it has been said that we could solve the problems of the cities, environmental pollution, transportation, and so on with wonderful on-the-shelf science—but, economic, political, sociological, and institutional barriers and legal barriers prove immovable.

Somebody has to find a way to make that transition. This council has been alert not only to the opportunities of science but to the feasibility of political process by which you can drag the scientific potential into the mainstream of political action. There is no higher body in the government that can do this than one at a cabinet level. What I am saying is that each initiative, when identified, consisted not only of a goal of social significance to the country but also of a plan and an allocation of responsibility to follow this plan through.

What are some of the programs? Let us consider the music that goes through this system. In its first year, the council turned its attention to the problem of world hunger. It is widely acknowledged that there are nearly one and a half billion people in the world with some degree of protein deficiency; to meet this deficiency, it has been estimated that there are at least five times as many fish in the sea as are being caught today among self-replenishing stocks. Given the problem, and given the solution, how does one connect the two together? A Food from the Sea program was developed, making use of a known but still immature technology of fish protein concentrates. The council developed a plan of year-by-year development of research and production capabilities, not just in the United States, but in other countries that suffer from hunger. In other words, we endeavored to develop a program that provides a continuous bridge between uncaught fish in the sea and unfed children on the land. This brought into the picture an agency which most would say had nothing to do with oceanography—the Agency for International Development, or AID. Within the federal array of agencies, AID already has a responsibility to wage the war on hunger. Having written out such a prescription, the council then transmitted a memorandum to the president proposing both program and responsible agency, and AID and the Bureau of Commercial Fisheries (BCF) under the Department of the Interior independently began to send signals to the Bureau of the Budget requesting necessary funds for implementation.

I have to interrupt this anecdote at this point to recall that like an audio system that needs to be plugged into some source of electricity, funds are required to meet any new goals. In this case, the Bureau of the Budget was alerted in time for them to consider a new and always unwanted claimant for funds and then to wrestle with the agencies in the usual budget process. It did not hurt to have the president approve this program, because the Bureau of the Budget works for the same man. When the boss says that this is something he wants, one can be sure of easing the inevitable budget problem. The point, however, is that an attempt was made not just to say, "More food from the sea," or "fish protein concentrate," or "We have to look at other species of oily fish," but to carry the system all the way through to what then became a proposal to the Congress last year. While only modest funds were involved, the policy to develop fish protein concentrates (FPC) and focus their potential outside our boundaries was adopted and the program began.

The next step was to help the lead agency succeed. The council staff assisted AID in recruiting individuals in the War on Hunger office who had the responsibility to run a "Food from the Sea" service. AID staff began to travel around the world, visiting a number of countries to determine their interest in some kind of bilateral FPC program, with the result one year later of an agreement with Chile as the first country in this program. Simultaneously, BCF was working to overcome several of its technical problems in refining the FPC extraction process.

But then we discovered another barrier: namely, the reluctance of the Food and Drug Administration to approve the sale of this product in the United States, which meant that we might be in a position of urging consumption of this product abroad when we had not approved it for our own domestic use. FDA is situated under HEW, and at one of our council meetings devoted to this impasse, Vice-President Humphrey just turned to Secretary John Gardner and said, "What are you going to do?" Under that direct question and cooperative response, the FDA, without having to change its standards, finally reversed its hostile stand, which frankly had been encouraged by some dairy interests who thought that protein sources from something other than dried milk might be serious competition.

A second interesting opportunity is one that President Johnson announced on 8 March 1968—a proposal for an International Decade of Ocean Exploration. This concept has a long history, but let me first describe the proposal and then give you a little background. The proposal that the president made was for all the nations of the world interested in the sea to cooperate and collaborate continuously and intensively in a major effort of exploration. The "decade" referred to is the decade of the seventies. There have been numerous expeditions before—such as the Indian Ocean Expedition and Tropical Atlantic—but these have been limited in scope: they start, they are conducted, and they stop. And there is a long interval in between. Sometimes it takes five years before the data get to scientists who may not have participated in the expedition and need the information. The concept in the president's proposal was to provide a continuing basis for the exploration of the sea by any individual groups or nations that are interested, providing them with a data exchange system using the newest possible technology. The concept has a number of new features: it anticipates intergovernmental as well as interscientific cooperation; a focus on resources and environmental preservation, as well as on pure science; and an attempt to set priorities by international planning, with a new banner to trigger funds, rather than piecing together studies that would have been conducted anyway.

Data are the commodity of cooperation, so an effort was made to find a management scheme for the handling of all kinds of oceanographic data, in data banks to which anyone can make a deposit and from which anybody can make a withdrawal.

As to decade planning, we have all kinds of glittering objectives, so that priorities are required. First of all, it is necessary to make clear that this proposal included not only expanded scientific study of the sea, but the delineation of resources. In other words, we are interested in finding out where the fish are and why, where the minerals, oil, and gas are, and what problems are emerging in pollution of the sea.

Second, the concept is one that would make use of as many existing instruments of international cooperation as possible, without inventing new ones. However, there is no single agency that satisfies the requirements for coordinating these multidisciplinary, multigoal activities. The Inter-Governmental Oceanographic Commission (IOC) under UNESCO was encouraged to get together with FAO and WMO in order to decide among themselves whether or not they could develop a coordinating mechanism. They indeed met, and it looks as if IOC will be given the coordinating responsibility.

Another related development occurred almost simultaneously. In August 1967, the United Nations was confronted with an imaginative proposal by the government of Malta of turning over to the United Nations all the mineral resources of the deep sea. This proposal immediately raised the legal question of who owns the seabed. The United States, in responding, was very happily not caught completely off balance, because it was planning to come into the United Nations with its own proposal, as a result of studies the council had urged the State Department to begin in March of that year. But the United States was still surprised at Malta's initiative. Ambassador Arthur Goldberg went to the United Nations in October with a three-part proposal: first, to re-enunciate a concept the government has adopted, that there be no colonial race with regard to the sovereignty of the deep seabed; second, that the United Nations establish a committee on the oceans; and third, that the nations of the world work together to find out what resources are in the oceans, because this important factual base is required when evaluating alternate legal regimes.

The United Nations subsequently adopted a resolution to establish an ad hoc committee on the oceans, which is potentially quite significant because of the parallel with outer space. Some of you may remember that the treaty on the banning of nuclear weapons in outer space, and other agreements of cooperation which were ratified only recently, came out of the United Nations Committee on Outer Space, which itself succeeded the ad hoc committee formed in 1958. That process took nine years. How long it will take in this case, if indeed the committee does act to develop directions of this kind, is uncertain. In any event, that Committee on the Oceans has been established, and it met for the first time in New York. It adjourned after having set up two subcommittees, one dealing with legal questions, one dealing with economic and technical questions. The Soviet Union also proposed that weapons of mass destruction be banned on the deep seabed.

At present, we still do not know what other countries think. The council and the State Department have been in touch on a bilateral basis with a number of these countries. With the decade concept in mind, I had the privilege of joining Vice-President Humphrey on his trip to Europe in 1967 and visited with marine science officials in six nations. I discovered that, first of all, other governments are just as complicated as our own and they have as many different ministries dealing with the oceans as we do. Recognizing this, in each country I requested a meeting of all the ministers involved in marine programs, and many of them sat down around the same table for the very first time to talk about the oceans. The only country which on its own had recognized that problem and had already done something about it was France. It has established a council in admitted imitation of ours, but with one important difference: their council has 20 percent of their marine science budget to redistribute to their operating agencies. That control over funds gives leverage, because they can get some new things going without cutting through the bureaucratic jungle as we do—making progress only by persuading other agencies to spend their money differently. Again, it is too soon to tell how successful they will be.

The purpose of citing these few examples is to give you an idea of the problems of getting things done in a democratic society where, in fact, you do not order government officials to do things. If you do, you are likely to be sabotaged by all the back doors that exist in Washington. So progress depends on another strategy, that of first formulating objectives that people think are important and in the national interest, and then trying to persuade people that it is in their own particular agency's interest to move in these directions. This is one of the less mysterious ways things happen in this government. I do not know whether anybody has ever estimated the energy it takes to get one dollar's worth of effort through the system; the overhead must be enormous. But even though the alternative of some other form of authoritarian government may be more efficient, it loses its respect for pluralism and individuality. So we try to figure out how to make this system work the best possible way.

It turns out that the council machinery is a model for some other problems, because, increasingly, issues are crossing agency lines. There is no longer a correspondence between the structure of government and the mission. The Department of Agriculture, through its PL-480 program, has an enormous influence on foreign policy, as, obviously, the Department of Defense does. The State Department is not uniquely our foreign policy agent. One can go down the line, mission by mission, or objective by objective and, except in a very few instances, it is difficult to find a correspondence between problems and organizational structure. One must consider whether you reorganize the government every time you have a new problem, or whether you find some kind of a government mechanism to deal with new issues across agency lines.

As suggested earlier, Washington has a penchant for measuring priorities by dollars. Actually, we each do this in our own families; although we might not like to admit it, the budget is a one-dimensional representation of an aggregate set of wants and needs of a family or society. And so, indeed, one needs to consider that to be the case in marine science affairs. One does not endeavor to increase a budget simply out of enthusiasm; it is necessary to have valid goals which require concurrence by all of the agency heads in a consensus, by the president, by the Bureau of the Budget, by the committees of the Congress, and so on. The fact of the matter is, then, that a case has to be made with strong justification, and one of the council's responsibilities has been to help make the case sufficiently persuasive to earn support.

The last thing that I wanted to comment on was the question of where we go from here, in terms of a study by this advisory commission. As you know, the act which set up the council also created the distinguished advisory commission, chaired by Dr. Julius Stratton. This advisory body was charged with preparing a report setting forth long-range goals and recommending an optimum government organization, with the implication that this council arrangement is not the ultimate. From where I sit, I would agree it is not the optimum. And this, then, immediately begins to suggest a wide variety of options in terms of how one wires these functions enunciated previously in a more effective way. You can see how many different agencies are involved.

It is going to be quite a challenge to analyze how to reorganize the government in order to get a better effect out of the components. The challenge, however, goes beyond simply restructuring the existing pieces, because there is the fundamental question of whether all the pieces are here. The most important missing piece deals with the developing and protecting of the resources of the ocean. As we look ahead, we see that our own fishing industry does not have the vitality and the viability that one needs in this kind of an enterprise. We do not have, nor does anyone, any mechanism for developing technology for deep-sea mining or mineral extraction. A very interesting question arises as to how important it is to deal with these international resources in a completely new way. And if you let your imagination run wild, you might even consider possibilities of a multigovernmental consortium.

Also in the picture is the question of the merchant marine. It turns out that the act establishing this council covers the question of transportation and commerce. Here again we find an industry that is losing ground, being artificially sustained through a subsidy program that was developed in 1936, unchanged since, where every interested party—whether it is the shipbuilders, the ship operators, the ship owners, the labor unions, or the Maritime Administration—is so rigid in its position that we have made no progress.

The council has necessarily had to look at these questions. According to the act, the council receives the report from the commission and is obliged to comment on it in forwarding it to the president. We endeavored to anticipate that review with a set of broad questions which all reduce to one: What is our nation's—and the world's—stake in the oceans? That may seem to be a simple set of words, but the fact is that this country has forgotten its maritime heritage; we have not really looked very hard at how important the oceans are, and the weakness of our ocean-related industries reflects this. At this time in history, as it seems to some of us, it is essential to ask another key question: How do the oceans contribute to the maintenance of world order? This goes well beyond the question of maintaining a navy: it goes into the issues of a maritime presence and of the opportunities that this decade of exploration will provide for people from many countries to work together toward a common goal. The question of seabed ownership that has to be solved at the United Nations is going to provide a similar opportunity. It is going to be quite a challenge to this country and to others to see whether we can meet this problem without reverting to some of our older nationalistic approaches. I am inclined to be optimistic, but it is going to take a concerted effort by a new group of players, new in the sense that this goes well beyond oceanography as a science to a field that puts people and their institutions in the picture. From the topics and processes mentioned here, it is clear that we are dealing with engineering, economics, foreign policy, banking, public administration, and international law. And that is really the substance of the subject, that a multidisciplinary and integrated approach is the key to the future of marine science affairs.

Index

Abalones, 66, 72
Acceptance index: equations of, 62
Agency for International Development (AID), 169, 173
Alaska: Cook Inlet, 26, 98; oil in, 97-98
Alaska pollack, 69
Albacore, 74
Algae: use of, 46. *See also* Multicellular plants
Amazon River, 4
American Bar Association (ABA), 137; report of, 142-43; resolution of, 153-54
American Mining Congress: "Resolution on Undersea Mineral Resources," 138-39, 143-44, 155; meeting of, 160; and international registry authority, 164. *See also* Wilkey, Malcolm R.
Anchovies: in composition of world fish catch, 65; as phytoplankton feeder, 68; as source of animal protein, 68; landings in 1948, 1958, 1966, 68; reason for large population of, 68; in Peru, 69; in California Current region, 85; scale deposits of, 86; economic significance in California of, 87; Peruvian fishery, 109
Antarctic krill, 70
Aquaculture, 63, 72
Aquatic resources. *See* Ocean resources
Artemia larvae, 62
Atomic Energy Commission, 169
Authigenic minerals: phosphorite, 22; glauconite, 22, 23

Baja California, 91
Barrier island: formation of, 16
Basses, 65

Bay of Fundy, 26
Beklemishev, Constantin: observations of, 58
Bering Sea, 98
Billfishes, 65, 69
Bluefin tuna, 69
Bonitos, 65, 66, 68, 70
Broad-grid bathymetric charting, 128
Bureau of Commercial Fisheries, 84, 173

Calanus, 56
California: marine margin problems of, 79-80; flood control projects in streams on beaches of, 83; and ability to control a fishery, 87; and marine investigations, 92
California Academy of Sciences, 92
California Current, 84, 86, 90
California Resources Agency, 92
California State Department of Fish and Game, 84
Capelin, 66, 69
Chesapeake Bay, 25
Chile: as first country in FPC program, 173
Chimaeras, 65
Clams, 66, 72
Clupeidae, 68
Clupeoids, 65, 66; abandonment of fisheries for, 73
Coal, 101
Coastal zones, 171
Coastline: defined, 11; dependence upon sea-level changes, 13; specific considerations of, 13; associated with a broad, low-relief coastal plain, 13, 15; cliffed, 13, 15, 19; prograding of, 16; transgres-

Coastline (*continued*)
sive sands, 19; nearshore zone, 20; geological history and characteristics of, 22
Cods, 65, 69, 73
Columbia River, 4, 25, 26, 108
Commission on Marine Science, Engineering and Resources (COMSER): creation of, 9; on marine mineral resources, 126, 131; conclusions of report of, 134-35; rejection of National Petroleum Council proposal, 140; on redefined continental shelf, 142; and U.S. policies, 145-46, 159-60; recommendation of international registry authority, 146-48, 151; recommendation of international fund, 148-49; and dispute settlement, 150; recommendation of "intermediate zone," 150, 152, 163; recommendation of national catch quotas for North Atlantic cod and haddock, 155-56; on leasing policies beyond redefined continental shelf, 165; recommendation for international, legal-political framework for exploration and development of mineral resources of high seas, 165-66; conclusions of, 166. *See also* International Panel
Commission to Study the Organization of Peace (CSOP): nineteenth report of, 156-57
Continental rises, 13
Continental shelf, 5; morphology of, 11; definition of, 11, 13, 99, 136, 137, 138, 139, 142, 145, 150; and glacial lowering of sea level, 19; surface configuration of, 20; types of sediments on, 20, 22; sand and gravel deposits on, 22; placer deposits on, 23, 127; as plant beds, 46; production rates of macroscopic benthic plants on, 46-47; maximizing animal protein yield of, 73; oil exploration on, 95-96; in North Sea, 99; lease boundaries for oil exploration, 99; energy from coal from, 102; oil and gas as energy sources from, 104; postwar fisheries expansion in waters of, 110; potential U.S. fisheries yield from, 114; predicted as area for future world oil production, 119; need for expanded national effort and basic technology toward exploration of, 131; views of international law on subsea areas beyond, 142; fixing outer limits of, 144; views on framework beyond limits of, 152; U.S. position on framework beyond limits of, 158
Continental Shelf Act of 1964, 99
Continental slope: defined, 11; aspect of, 13; in California, 83
Continental terrace, 11; defined, 13
Convention on the Continental Shelf (U.N.), 13, 141, 145-46, 149, 151
Convention on Fishing and Conservation of the Living Resources of the High Seas, 150
Copepod, 52, 53, 59, 60, 88; as food of herring, 68
Core sample: in oil exploration, 96
Council of Economic Advisors, 169
Crab, 66
Creole Field (La.), 95
Crude oil, 98
Crustaceans, 66, 69

Declaration of Santiago, 141
Deep-sea smelts, 68
Deep-sea trench, 13
Delta, 15
Demersal fishes, 45; fisheries of N. Atlantic, 110
Department of Agriculture: and PL-480 program, 176
Department of Defense, 176
Department of Health, Education and Welfare (HEW), 174
Department of the Interior, 162, 173
Department of State, 176
Desalination: in California, 80-82; economic feasibility of, 107; economics of, 108; flash distillation process, 108; problem of concentration of brine, 108-9
Diatom, 52
Dunaliella, 56

"Ecological efficiency," 60
Ecosystem, 40
Estuaries: embayed, 15; defined, 25; bar-built, 25; coastal plain, 25; characteristics of, 26; effect of density stratification on, 27; salinity of, 27; fluctuations associated with turbulence, 27-28; measurements of properties of, 28; "ensemble average," 28; currents and density regime, 29; dynamics of, 29; horizontal salinity gradient, 29; tidal dynamics of, 29; tidal regime, 29; types of, 29-30; fjord, 29-30; James River, 32

Estuary tide: progressive wave, 26; standing wave, 26
Euphausids, 68
Excreta: measurement of, 53-54

Feeding rate, 56-58
Filtration rate: description of, 56-58
Finlay, Luke W., 154
Fish: new uses for, 69; nutritional value of animal protein from, 69; preservation of, 69; carnivorous, 69; smaller, for expanded food production, 70; potential use of, relative to size, 70; U.S. consumption of, 71; nutritional and economic value of, 71; U.S. industry, 71-72; Russian industry, 72
Fisheries: composition of world catch in 1948 and 1966, 65; growth of, 66, 109; future development of, 69, 111, 112-13; and public policy, 73-74; state management of, 74; in California, 83-84, 87, 88; as component of marine resources, 106, 109; importance of world catch, 110, 111; present world status of, 113; in U.S., 113-14, 115; harvesting concept, 115; jurisdiction of, 115; overexploitation of, 117; need for management program on international and national level, 117
Fish meal, 66; new use for fish, 69; world production in 1948, 1958, 1967, 69
Fish protein concentrates (FPC), 173
Fjords: defined, 25
Flat fish, 69, 73
Flounders, 65, 66, 72
Food: concept of, 52-53; measurement of, 53-54
Food and Agriculture Organization (FAO), 65; world fish statistics classified by, 65; figures on annual world harvest of seaweeds and kelps, 67; report on tuna, 109-10; mentioned, 170, 175
Food and Drug Administration, 174
Food chain. *See* Marine food chain
Food from the Sea program, 173
Foraminifera, 22, 56
France: establishment of a marine science council in, 176
Fresh water: as component of marine resources, 106; from interbasin transfers, 107
Fur seal, 73

Gadoids, 66
Glauconite, 22, 23

Glomar Challenger, 144
"Grandfather rights," 160

Haddocks, 65, 73
Hakes, 65, 73, 86
Halibuts, 65, 66, 71, 74; conservation of, in Pacific Northwest, 73
Halophytic plants, 81-82
Hard mineral resources: future of, 123; demand for, from marine environment, 122-24; national policy of, 125; source of raw materials, 125; types of, 126; in submerged placer deposits, 126; need for further analysis of potential of, 128; role of government in development of, 130-31
Herrings, 65, 68
High Island Field (Gulf of Mexico), 95
Holocene transgression, 19
Hood Canal, 25
Hope Valley, 19
Hudson Bay, 98

Indian Ocean Expedition, 174
Inter-Governmental Oceanographic Commission (IOC), 175
Intermediate zone, 150-52, 162-65
International Convention on the Continental Shelf, 135, 152
International Court of Justice, 150
International fund, 148-49
International Law Association (ILA): Committee on Deep Sea Mineral Resources, 138; American Branch Interim Report, 154
International Panel (of COMSER), 134, 139; report of, 138, 147, 149, 150, 152, 156, 158
International registry authority, 146-48, 149, 162
Isobath: twenty-five-hundred-meter, 150-51; two-hundred-meter, 136, 137, 140, 142, 160, 161

Jacks, 65
Japan, 89
Johnson, Lyndon B., 161, 166, 174

Kelps, 67
Krill, 68

Lagoon, 16
Lobster, 66

Mackerels, 65, 87
Malta Resolution, 145, 175

Manganese nodules, 128, 129, 156
Maracaibo, Lake, 94
Marine aquaculture. *See* Aquaculture
Marine biology, 50
Marine food chain, 49-50, 53; compared with land-based food web, 50-51; mass-balance equation of, 54-55; measured rates of, 55; rates of feeding, 56; ratio of filtration rate to feeding rate, 57; efficiency of, 58, 59, 62-63; transfer efficiency, 60; efficiencies of processes, 60; benefit to man, 68. *See also* Trophodynamics
Marine gravity surveys, 97
Marine resources. *See* Ocean resources
Marine Resources and Engineering Development Act of 1966, 9, 144, 160, 163, 168
Median line principle, 99
Menhaden, 68
Merchant marine, 177
Microbial action: as potential energy source, 103-4
Mineral resources. *See* Hard mineral resources
Mining rights: on ocean floor, 143
Mission Bay, Calif., 83
Mississippi River, 25, 32
Molluscs, 66, 69
Monterey, Calif., 88
Mullets, 65
Multicellular plants (kelps, seaweeds, algae), 67
Mussels, 66, 72
Myctophids, 68

National Aeronautics and Space Administration, 169
National Commission on Marine Science, Engineering and Resources. *See* Commission on Marine Science, Engineering and Resources
National Council on Marine Resources and Engineering Development, 9, 169, 172
National Marine Fisheries Service, 115, 116
National Petroleum Council (NPC), 137; on offshore petroleum recovery, 144, 152; Interim Report, 144, 153
National Science Foundation, 144, 169
National Sea Grant College and Program Act of 1966, 9
1958 Geneva Convention on the Law of the Sea: and definition of continental shelf, 99; accompanied by "Optional Protocol Concerning the Compulsory Settlement of Disputes," 150
Nixon, Richard M., 161
Norway, 25
Nuclear energy, 102
Nutrient matter: subject to gravity, 40

Ocean: use of, 3, 7, 8-10; potential for man, 5; need for predictive procedures, 6; need for techniques of modification and control, 6; potential of, 10, 67-68, 102-4, 126; common property problem in, 116
Ocean basins, 11
Oceanographic Commission of Washington: creation of, 9
Ocean perches, 69
Ocean resources: defined, 3; condition of, 4; as replacement of scarce land resources, 6; viewed as uses of the environment by man, 7; living aquatic, 64-66; components, 107-11; development of, 106, 107, 144-45, 159, 177; need for protein from, 111; and COMSER, 134, 135; proposed course of interim action, 159
Ocean Science and Technology Advisory Committee (OSTAC): Mining Panel, 155, 156, 160
Octopus, 66
Office of Science and Technology, 169
Offshore bar, 16
Offshore oil exploration: history of, 94-96, 118; economic investment in, 95, 119; drilling methods, 97; expansion of, 98; legal, political, and economic aspects of, 98-101, 120; in North Sea, 99, 100; leasing regulations and agreements, 100; environmental problems, 100-1; as major component of marine resources in U.S., 119; world production of, 119; common-property dilemma, 120; production and import restrictions, 121; "industry in being" concept, 121; development correlation with onshore production, 122; Santa Barbara channel leases, 122; need for reassessment of American policy on, 122
Oil: recovery of, 7; aerial magnetometer surveys, 97; gravity and magnetic surveys, 97; supply prospects of, 118-19; production, 119; from shale and tar sands, 119; extraction of, from seabed, 119-20. *See also* Crude oil; Offshore oil exploration

Oregon Submerged Land Act, 95
Organic debris, 91
Outer Continental Shelf Lands Act, 161, 162
Oysters, 66, 72

Pacific mackerel, 87
Passamaquoddy Bay (Maine), 103
Percomorphs, 69
Persian Gulf, 142
Peru, 89
Petroleum. *See* Oil
Phosphorite, 22, 127
Photosynthesis, 52
Phytoplankton, 38-40 *passim*, 43-47 *passim*, 50, 60, 61, 63, 70; seasonal cycle of, 43, 45; production of, 43-45; world productivity, 62; harvesting of, 67-68
Pilchard, 68
Placer deposits: on continental shelves, 23, 127; submerged, 126
Plankton: herbivorous, 39; animal, 50; measured rates of, 55-56; trophodynamics of, 58; unicellular plants, 67; mentioned, 70
Pleurencodes (red crab), 68
Point Loma, Calif., 19
President's Science Advisory Committee: Panel on Oceanography (1966), 6
Pteropod (*Limacina helicina*), 90
Public Land Law Review Commission, 131
Puget Sound, 26, 88

Rance Estuary, 103
Rays, 65
Red crab (*Pleurencodes*), 68
Red fishes, 65
Resources. *See* Ocean resources
Rhincalanus nasutus, 62
Rhone River, 103
River runoff, 45-46

Salmon, 66, 69, 71, 74
Salt domes: in Gulf of Mexico, 128
San Diego, 88
Sannich Inlet (Vancouver Island), 91-92
Santa Barbara Basin, 89, 91
Sardinella, 68
Sardines, 65; purse-seine fishery for, 84; scale deposits, 86
Scripps Institution of Oceanography, 80, 84
Scuba diving, 96
Sea urchins: chemical control of, 67

Sea water: culture of ordinary crop plants in, 82
Seaweeds, 67
Seiches: estuary fluctuations associated with, 27
Sharks, 65
Shrimp, 66, 68, 72
Sigsbee Knolls (Gulf of Mexico), 144
Silver Bay (Alaska), 25, 32
Skipjacks, 66
Smelt, 66
Smithsonian Institution, 169
Soles, 65, 66
"Sovereign rights," 139, 142
Sperm whales, 40
Sprat, 68
Squids, 66
Stone, Oliver L., 138
Strand plain: composition of, 16
Stratton, Dr. Julius, 177
Stratton Commission, 171
Subsea resources. *See* Ocean resources
Substrate rocks, 127-28
Summerland Field (Calif.), 94

Tar sands and oil shales, 101
"Territorial sovereignty," 142
Thermocline, 40, 43
Thread herring, 68
Tidal currents: effect of, on salt transport, 26; energy provided by, 27
Tidal energy, 103
Tidal marsh, 16
Tidal motion, 103, 107
Tidal river, 25
Tides: bathymetric effects of, 26; Puget Sound, 26; Columbia River, 26; Bay of Fundy, 26; Cook Inlet, 26; currents, 26; and internal waves, 26
Tin, 130
Transfer efficiency, 60
Trophic level, 52
Trophodynamics, 49, 53, 58, 60, 63, 68. *See also* Marine food chain
Tropical Atlantic, 174
Trout, 66
Truman Proclamation of 1945, 139, 141, 150
Tuna, 65, 66, 69, 71; development of fishery of, 84, 88

Union of Soviet Socialist Republics, 175
United Nations: ad hoc committee on uses of seabed, 136; Political Committee, 145; and redefinition of continental shelf, 145; on exploration and

United Nations (*continued*)
exploitation of mineral resources in sea, 145; and international fund, 148; Development Program, 148; and international registry authority, 157; and administration of use of resources in sea, 165; policy planning for marine sciences, 170; Committee on the Oceans, 175; Committee on Outer Space, 175. *See also* Convention on the Continental Shelf; United Nations Educational, Scientific, and Cultural Organization

United Nations Educational, Scientific, and Cultural Organization (UNESCO), 170, 175

United States: and "Optional Protocol," 150; proposals to U.N. on development of minerals in ocean, 158-59, 175; authorizing exploitation of minerals in seabed, 159-60; and registration of claims with international registry authority, 161-62; and intermediate zone, 163; and claims beyond intermediate zone, 164

U.S. Bureau of Mines, 118

U.S. Coast and Geodetic Survey, 46

University of Washington: Division of Marine Resources, 9-10

Upwelling, 40

Varved sediments, 86, 89

Vertical mixing: mechanism of, 40; by convection, 43

War on Hunger office: "Food from the Sea" service, 173

Water power: as source of energy, 102

Water production. *See* Desalination

Waves: internal, 103; surface, 103

Weather prediction, 83

Wilkey, Malcolm R.: and resolution of American Mining Congress, 139, 144, 155; on international registry authority, 164

Wisconsin glacial advance, 18

World Bank, 148

World Meteorological Organization (WMO), 170, 175

Yellowfin tuna, 70

Yellowtail, 72

Zooplankton 38-40 *passim*, 57; herbivorous, efficiency of, 39, 43, 50, 54, 61, 63

DATE

11. 12. '87